RTTY/AMTOR Companion

By Steve Ford, WB8IMY

Published by: **The American Radio Relay League**
225 Main Street, Newington, CT 06111

Copyright © 1993 by

The American Radio Relay League, Inc.

Copyright secured under the Pan-American Convention

International Copyright secured

This work is publication No. 165 of the Radio Amateur's Library, published by the League. All rights reserved. No part of this work may be reproduced in any form except by written permission of the publisher. All rights of translation are reserved.

Printed in USA

Quedan reservados todos los derechos

First Edition
First Printing

$8.00 in USA

ISBN: 0-87259-409-2

Foreword

Picture how text is sent via radio. On the VHF, UHF and microwave bands, it's relatively easy. On the HF bands, however, it's another matter entirely! Not only are the digital signals squeezed into smaller chunks of spectrum, they have to contend with fickle propagation and a much higher level of interference.

Immediately after World War II, amateurs began using surplus radio teletype, or RTTY gear. For decades thereafter, the musical warblings of RTTY stations could be heard as they conversed over hundreds and thousands of miles. Fading and interference caused errors in reception, and the equipment was big and noisy but, most of the time, RTTY worked. In the early 1980's, the era of the personal computer brought RTTY to the desktop and added a new word to the Amateur Radio lexicon: AMTOR. The synchronized, chirping rhythms of AMTOR signaled the beginning of virtually error-free digital communications on the HF bands.

RTTY and AMTOR remain with us in the '90s—and they are as enjoyable and fascinating as ever. New modes such as CLOVER and PacTOR have joined their ranks, providing spectacular improvements in the quality of digital transmission and reception.

Steve Ford, WB8IMY, is about to take you on a tour of these intriguing modes. You'll learn how to build and operate your own RTTY and AMTOR stations. You'll also be introduced to PacTOR and CLOVER. If you're new to Amateur Radio, I encourage you to explore HF digital communications. You're in for a lot of excitement! If you're a veteran in need of fresh opportunities, this book points the way.

The HF digital modes have a great deal to offer: relaxed, conversational communications; challenging contests; rare

DX contacts; public service operations and much more. As you'll discover in *Your RTTY/AMTOR Companion*, you don't need to be a digital expert to get started. All that's required is the capacity for awe and wonder!

    David Sumner, K1ZZ
    Executive Vice President
    March 1993

Acknowledgments

Don't let the cover of this book fool you. It takes more than one individual to create *any* book. In the case of *Your RTTY/AMTOR Companion*, there is an army of behind-the-scenes people who spent a great deal of time and effort to make this book a reality. If I listed everyone, we'd have to add an entire page—maybe two!

Instead, I'd like to call your attention to several individuals whose contributions were especially critical:

Bill Henry, K9GWT: I had the pleasure to work with Bill on a couple of RTTY and AMTOR articles that appeared in *QST* magazine in 1992. Bill's company, HAL Communications, is helping to pioneer what may be one of the most revolutionary HF digital modes in decades: CLOVER. Much of the CLOVER information in Chapter 5 was provided by Bill Henry. What I've *learned* from Bill is scattered throughout the book!

George Hitz, W1DA: George was my "fact checker" for this project. In the true spirit of communications technology, we swapped chapters and comments via electronic mail. During the writing process, we never spoke to each other directly. Our keyboards did the talking! George's thorough research and insightful comments enrich each chapter.

Dale Sinner, W6IWO: Dale is the publisher of the *RTTY Journal* and he was gracious enough—and patient enough—to answer all my nagging questions. The contest information you'll find in Chapter 6, for example, is courtesy of Dale.

Cover photo: Kirk Kleinschmidt, NTØZ

Contents

Foreword

1 What's that Racket in my Radio?

2 Building Your Own RTTY/ AMTOR Station

3 Your First RTTY Conversation

4 Time to Start Chirping with AMTOR!

5 Exploring PacTOR and CLOVER

6 RTTY/AMTOR Contesting

Glossary

RTTY/AMTOR Resource Guide

About the ARRL

Index

CHAPTER 1
What's That Racket In My Radio?

If you've spent time exploring the Amateur Radio HF bands, I'll bet you've heard some pretty strange stuff. I'm not talking about the occasional operator who appears to be a few sandwiches short of a picnic, so to speak. No, I'm referring to the odd sounds that you'll find just above the CW portions of the bands.

Let's take a trip through this strange landscape. Do you hear warbling, musical signals? They've been around for decades and veteran amateurs recognize them immediately as *radioteletype*. Although there are many radioteletype codes, the ham custom is to refer to *Baudot* or *Murray* radioteletype as *RTTY* (pronounced "ritty"). Twist the dial a bit further and you're likely to hear a chorus of electronic crickets. These are the chirping dialogs of *AMTOR* (*AM*ateur *T*eleprinting *O*ver *R*adio) stations. Keep hunting and you may also hear the unusual sounds of *PacTOR* or *CLOVER*. And what about those raspy, high-pitched bursts? Those are the unmistakable signatures of *packet*.

These are the primary HF *digital* modes. They're called digital modes because the communication involves an exchange of digital data between one station and another. In the case of *AFSK* RTTY, for example, letters typed on a keyboard are translated into data by a computer or data terminal. Another device, usually a *multimode communications processor* (*MCP*) or a *terminal unit* (*TU*), accepts the data and converts it to whatever encoded audio tones are required. (We'll discuss both of these devices in Chapter 2.) The tones are sent to the transmitter and away they go! At the receiving end the same process occurs in reverse: The tones are translated back into data and displayed as text on a computer or terminal screen.

Who Cares?

Why should you care about HF digital modes? If you have something to say, isn't it easier to pick up a microphone or grab a CW key? The answer depends on how effectively you want to communicate.

With the exception of RTTY, all digital communications modes include some form of error detection. This means the text from your station will arrive at its destination *without errors*. It will be received exactly as you sent it—complete with typographical, spelling and grammar glitches. (You can't expect your communications system to correct those problems, too!)

Can you say the same thing for a phone or CW conversation? Well . . . yes, as long as conditions are decent. Even when noise and interference are severe, the human brain has a remarkable ability to recover meaningful information. However, your mind also has a tendency to use imagination to fill some of the gaps. You may *think* you understood everything correctly, but did you?

A microprocessor-based system isn't burdened with imagination. It receives the data correctly, or it doesn't—there's no room for random speculation. Its attention span

So What Can I do with RTTY and AMTOR?

I thought you'd never ask! Here's a partial list...

- ❏ Enjoy conversations with amateurs throughout the world. The somewhat slower pace of RTTY and AMTOR allows you to compose your thoughts and express yourself clearly. As a result, you'll find that RTTY and AMTOR conversations are friendly and relaxing.
- ❏ Work rare DX. Many DXpeditions operate RTTY and AMTOR in addition to SSB and CW. By using your RTTY/AMTOR capability, you'll have an extra edge on the competition!
- ❏ Access RTTY and AMTOR mailboxes. By connecting to a mailbox, you can read bulletins, drop off messages for other operators or read messages that have been left for you.
- ❏ Use AMTOR *APLink* systems. An APLink system is your gateway to the packet radio network. You can post messages on an APLink and they'll be forwarded to the VHF/UHF packet network for delivery to hams at their local packet bulletin boards. Packet-active hams can also send messages to you via the APLink system.
- ❏ Perform public-service work. In emergency situations, RTTY and AMTOR are often in the spotlight. If disaster strikes in a distant part of the world, RTTY and AMTOR operators become communications lifelines. For example, injury lists and detailed requests for aid are easily handled by RTTY and AMTOR stations. The National Traffic System also depends on RTTY and AMTOR to move vital messages throughout the world.
- ❏ Participate in contests that *really* test your skills as a RTTY or AMTOR operator!

never wanders, no matter how long it has been operating. It proceeds with its instructions diligently and patiently until you tell it otherwise. As a result, the digital modes are highly effective and reliable. With recent developments in digital signal processing (DSP) and innovations such as PacTOR and CLOVER, hams are swapping digital data even under very poor signal conditions!

Although there are obvious practical benefits to using HF digital modes, you can't forget the greatest benefit of all: they're *fun*! Operating the HF digital modes is a unique experience. It's always thrilling to send a RTTY CQ and watch your screen as a stranger—across the continent or half a world away—responds to your call. As the text appears, you read it eagerly, the same way you'd read the opening pages of a good novel. A few years ago I enjoyed a fascinating RTTY conversation with a commercial airline pilot. I kept sending one question after another (How much thrust does an engine

The more the better when it comes to contesting! Jon, KB9ATR (third from the left) did a multioperator effort in the 1992 ARRL RTTY Roundup to get more people interested in digital communications.

on a Boeing 767 develop?) and he patiently answered each one. I printed the text of the conversation and I still have it today.

AMTOR is just as enjoyable. Even though I've made hundreds of AMTOR contacts, I still get a tingle whenever I establish a *link* to another station. My transceiver begins switching rapidly back and forth from transmit to receive. The room is filled with a rhythmic chirp-chirp melody as the data exchange begins. The letters creep across the screen in fits and starts. A conversation has begun!

Since RTTY and AMTOR are two of the most popular HF digital modes in use today, they are the primary focus of this book. I'll also introduce you to PacTOR and CLOVER—two related modes that are rapidly gaining a foothold. HF packet is extremely popular, but it requires a separate discussion that goes beyond the scope of this book. If you'd like to learn more about HF packet, I recommend the ARRL's *Your Packet Companion* or the *ARRL Operating Manual*.

A Short History Lesson

In the sense that we interpret the word "digital" today, RTTY has the distinction of being the granddaddy of HF digital communications. RTTY dates back to World War II when the military began connecting mechanical teletype machines to HF radios.

At first they tried simple on/off keying to send text, but that didn't work very well. The receiving equipment couldn't always tell the difference between a signal and a burst of noise. After some further experimentation, the designers switched to *frequency shift keying (FSK)*. This approach used two specific tones to indicate the on/off (*MARK/SPACE*) signals. FSK was a success and RTTY as we know it today was born!

Hams adopted RTTY after the war ended and by the

early '50s it was a well-established mode. Initially it found a home on VHF, but later became more popular on the HF bands. (RTTY is seldom heard on the VHF bands today.)

For several decades, hams relied on surplus teletype machines for their RTTY stations. These mechanical monsters were slow, noisy and often dirty (they had a nasty tendency to drip oil on the floor!). Operators had to read the text on paper as it was printed. The keyboard was a bit unusual, mainly due to the nature of the Baudot code.

With Baudot, all letters are capitalized (upper case). All numbers from 0 through 9 are available along with some limited punctuation. To send numbers or punctuation, a special *FIGS* character must be sent *first*. To return to alphabetical letters, a *LTRS* character must be sent. On the original teleprinters you had to press the FIGS key whenever you wanted to send numbers or punctuation. To return to alphabetical letters, you had to press the LTRS key (see Fig 1-1). You can imagine how difficult it must have been to master the keyboards of those old machines!

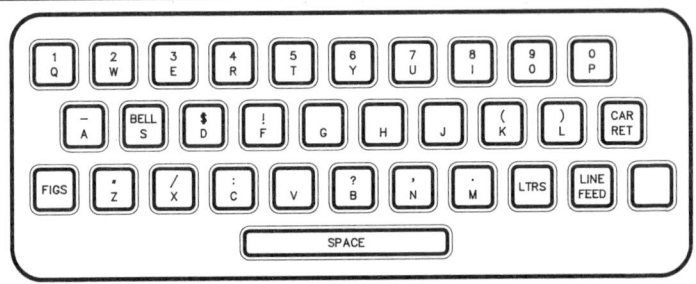

Fig 1-1—The old teleprinters featured a keyboard layout similar to the one shown here. Notice the FIGS key in the lower left corner. You had to press this key before sending punctuation or numbers. Pressing the LTRS key returned you to the "letters" mode. Although these vintage units have all but disappeared, the need to shift from FIGS to LTRS and back again remains. This is now handled automatically by software.

Al, RC2AZ, is an active RTTY operator in Russia. He built virtually all the equipment shown in this photograph!

Some ancient teletype machines remain, but most RTTY enthusiasts rely on computerized systems. The Baudot code is still the same, however, and the FIGS/LTRS shift is still required. No need to worry about your typing skills, though. The FIGS/LTRS shift is handled automatically. All you have to do is type and your software and hardware will take care of everything else. Rather than reading the text on flowing sheets of paper, you'll see it on your monitor screen.

RTTY has fulfilled its promise of transmitting the written word throughout the world. When band conditions are good and signals are optimal, RTTY is efficient and accurate. But what happens when conditions are less than ideal? That's when RTTY shows its weak side. Interference from other transmitters, fading and electrical noise cause errors in RTTY communications. Mild interference will cause a few letters to be deleted here and there. Severe interference can turn the entire text to gibberish!

Commercial maritime communication systems relied on RTTY for decades, but the inability to detect errors was a persistent problem. It could even have life-threatening consequences. What if a maritime weather service tried to alert ships of an approaching storm? The RTTY-transmitted warning had to be repeated over and over to give the ships a decent chance of copying the entire message.

The pressure to create a more reliable teletype system lead to the development of *TOR* (Teleprinting Over Radio), commonly known today as *SITOR* (Simplex Teleprinting Over Radio). Instead of sending the text in one long transmission, the TOR method sends only a few characters at a time. The receiving station checks for errors using a bit-ratio-checking scheme. If all characters are received error-free, the receiving station sends an *acknowledgment* or *ACK* signal and the next few characters are transmitted. If an error is detected, a *nonacknowledgment* or *NAK* signal is sent. This tells the transmitting station to repeat the characters. The result is digital communication *without* errors—a major improvement over previous RTTY systems. The rapid error-checking dialog—known as *Mode A* or *ARQ*—creates the distinctive chirping sounds associated with SITOR communications.

In the early 1980s, the Federal Communications Commission approved SITOR techniques for Amateur Radio use. Peter Martinez, G3PLX, adapted SITOR coding and developed AMTOR. AMTOR was tailor-made for the personal-computer era and it quickly became popular. (The advent of PCs also made *ASCII* RTTY possible using a *complete* character set including upper- and lower-case letters. However, its use remains somewhat limited today because of its lack of error detection.)

AMTOR uses the same limited character set as RTTY, but the coding is different. AMTOR sends 7 data bits per character instead of 5 used by RTTY. Like RTTY, AMTOR can send only upper-case letters. (This is changing, however. See the sidebar, "Upper/Lower Case AMTOR.") AMTOR shares another aspect with RTTY: it sends data by using frequency-shift keying.

You Take the High Tone and I'll Take the Low Tone

Understanding the nature of frequency shift keying, or FSK, is important if you're going to be an informed RTTY or AMTOR operator. When it comes to RTTY and AMTOR, FSK is everything!

Let's start with data. I'm sure you've heard that the fundamental language of all computers is binary *machine code*. In a binary-number system, you're only dealing with 0s and 1s. This is a natural situation for a computer since it's

---

Upper/Lower-Case AMTOR

AMTOR evolved from commercial TOR and SITOR, which in turn evolved from Baudot radioteletype. One of the handicaps of this evolution was that AMTOR could send only upper-case letters and limited punctuation. Peter Martinez, G3PLX, and Victor Poor, W5SMM, have created an extension of AMTOR, using the *null* code as an upper/lower-case shift signal for receiving stations.

By using upper/lower-case AMTOR, additional punctuation symbols are transmitted as well. In fact, the W5SMM version of APLink includes all of the standard punctuation symbols. For this reason, some operators refer to mixed-case AMTOR as *ASCII AMTOR*. It really isn't the full ASCII character set, but it's close.

At the time this book went to press, only the HAL PCI-3000 and AMT series controllers supported upper/lower-case AMTOR. No doubt other manufacturers will incorporate this feature as it becomes more popular. Stations using upper/lower-case AMTOR—including APLink BBSs—are fully compatible with upper-case-only users.

comprised of a multitude of solid-state *logic* switches that can only be *on* or *off* ("high" or "low"). So, an "on" condition represents a binary 1 while an "off" condition represents a binary 0.

If you use wires to connect two computers, the on/off voltage states are communicated from one machine to another easily. But let's make the situation more complicated and move the computers several hundred miles apart. Now what are you going to do? Radio seems like a natural choice, but you can't send high/low voltage states over the air. . . or can you?

What if you translated the changing voltages to changing *tones*? You could use 2,125 Hz to represent a binary 1 and 2,295 Hz to represent a binary 0. Feed those tones to the audio input of an SSB transceiver operating on lower sideband, for example, and they'll be transmitted as signals at specific points below the *suppressed carrier frequency*. The 2,125-Hz tone will create a signal 2,215 Hz below the suppressed carrier. The 2,295-Hz tone will create a signal 2,295 Hz below the suppressed carrier. Subtract the frequency of the high tone from the low tone and you get 170 Hz. In other words, the tones *shift* 170 Hz to represent a 1 or 0. Shifting voltages have become shifting tone frequencies! As we discussed previously, the tone that represents a binary 1 is called the MARK. The tone that represents a binary 0 is called the SPACE (see Fig 1-2).

At the receiving end of the path, you'll need to convert the tones back into binary high/low voltage states. FSK demodulators are designed with audio filters to detect the MARK and SPACE tones and produce corresponding data pulses. Feed those data pulses to a computer running *terminal* software and—trumpet fanfare please!—text appears on the screen.

RTTY and AMTOR use frequency shift keying to pass

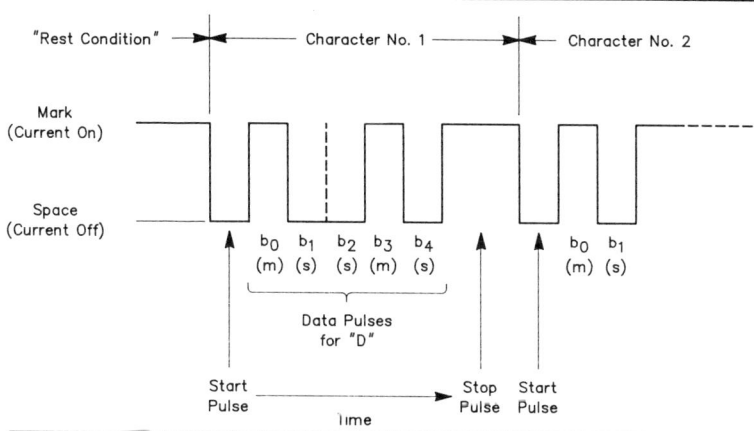

Fig 1-2—This is a diagram of RTTY MARK and SPACE signals as the letter "D" is sent. A start pulse begins the character, followed by the 5 bits that define it (b_0 through b_4). By looking at the bits, you can see that "D" is MARK-SPACE-SPACE-MARK-SPACE. A stop pulse signals the end of the character and another start pulse begins a new character. These MARK and SPACE pulses appear as shifting audio tones at the receiver. They are fed to a multimode communications processor (MCP) or a terminal unit (TU) to be decoded back into data and displayed as text.

information from one station to another. A 170-Hz shift between the MARK and SPACE tones is the Amateur Radio standard. A few hams use an 850-Hz shift, an artifact from the early VHF RTTY days. Commercial SITOR stations often use a 425-Hz shift. Most MCPs and TUs available today offer all three shifts. However, many MCPs fudge the 170-Hz shift to accommodate HF packet as well as RTTY and AMTOR (see the sidebar "Shifty MCPs").

FSK or AFSK?

In our previous example, the shifting tones were supplied to the transmitter via an audio input jack. Technically speaking, this is *audio* FSK or AFSK. There is another

> **Shifty MCPs**
>
> Some multimode communications processors (MCPs) use a 200-Hz shift for RTTY operation (mark = 2110 Hz, space = 2310 Hz). These tones are used for HF packet radio as well. They are compatible with standard 170-Hz shift tones (2125/2295 Hz) since the center frequency (2210 Hz) is the same. In weak-signal conditions, using a 200-Hz shift places you at a disadvantage. Both you and the other station (possibly using 170-Hz shift) must resort to *straddle tuning* to achieve true transceive operation. This is tricky and often results in less than optimum performance.
>
> It's possible to obtain optimum RTTY performance by retuning the MCP receive filters and transmitter tones to match the 2125/2295-Hz standard. While this technique offers a noticeable improvement, you should *not* attempt the adjustment unless you are thoroughly familiar with your MCP and possess all the necessary test equipment. (Most MCP manuals *do not* describe this procedure.) A botched realignment will probably void your MCP warranty!

method known as *direct* FSK. What's the difference? If you're using *direct* FSK, the data pulses from your MCP or TU are *not* converted to MARK and SPACE tones. Instead, they're applied to the FSK input of an SSB transmitter where they shift the frequency of the master oscillator up and down (see Fig 1-3).

So which is better—direct FSK or AFSK? Some hams believe that FSK is a "purer" method of transmitting RTTY and AMTOR because it minimizes distortion and harmonics. Other hams argue that if the transmitter isn't overmodulated, AFSK transmissions are just as pure. You'll find large numbers of amateurs using one method or the other, although the majority seem to favor the AFSK approach because most transceivers do not feature an FSK input. The important thing to remember is that whether you use AFSK or direct FSK, the result at the receiving station is the *same*.

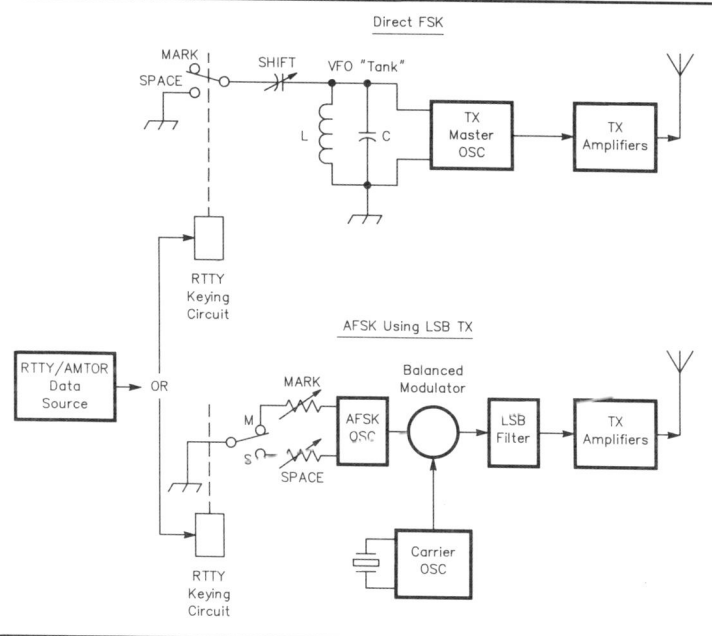

Fig 1-3—RTTY and AMTOR are transmitted by direct frequency shift keying (FSK) or *audio* frequency shift keying (AFSK). The upper portion of the diagram illustrates direct FSK. The MARK and SPACE data is applied to a keying circuit that "pulls" the master oscillator in the SSB transmitter, causing it to shift its frequency. In this example, the SHIFT capacitor is added in parallel to the VFO whenever a SPACE pulse is applied.

In the lower portion of the diagram, the MARK and SPACE data is applied instead to an AFSK oscillator (usually this oscillator is inside the MCP or TU). The oscillator generates tones that are fed to audio amplifiers within the SSB transceiver and, ultimately, to its balanced modulator. Regardless of whether you use AFSK or FSK, the result at the receiving station is the same.

If you already own an HF SSB transceiver, you're well on the way to creating your digital station. The next item you'll need is a terminal program, an MCP (or TU) and a computer or data terminal. As you'll see in the following chapter, there are many ways you can keep costs to a

What's That Racket In My Radio? 1-13

minimum. And if you don't own an HF transceiver, shop around for an inexpensive used radio.

Turn the page . . . and let's start building your HF digital communications station!

CHAPTER 2
Building Your RTTY/AMTOR Station

I'll bet you thought this was going to be one of the most difficult chapters in the book. Well, don't let the idea of building a RTTY/AMTOR station intimidate you. Depending on how complex you want to get, it can be as simple as attaching a few cables and making a couple of adjustments.

Take a look at Fig 2-1. This is a diagram of a basic

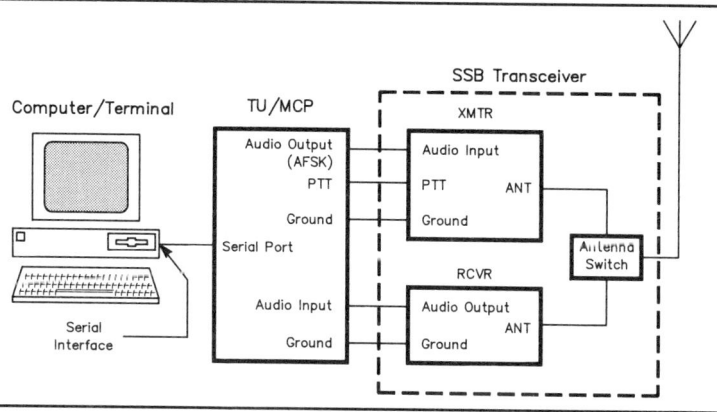

Fig 2-1—All you need is a computer or data terminal, a multimode communications processor or terminal unit and an SSB transceiver. As you'll read in this chapter, used equipment is often adequate for RTTY and AMTOR. In some cases, you don't even need a computer!

RTTY/AMTOR station. As you can see, there are three critical areas: the computer, the radio and the MCP or TU. We'll start from the left side of the diagram and work our way across . . . beginning with the computer.

Does it Compute?

If you're like many hams, you probably have a computer of some kind at your station. (If you don't, bear with me. I have some suggestions for you, too.) Computers are ideal for any digital mode. When running the proper software, they become the perfect interface between you and your equipment.

Does it matter what type of computer you're using? Not really. As long as it can communicate with the MCP or TU, you're in business. To make this possible, it must be equipped with an EIA-232-E *serial* port (more commonly called an RS-232 port), or a TTL (transistor-transistor logic) interface.

GØARF just completed a successful ARRL RTTY Roundup contest despite gale-force winds and two power outages.

Most MCPs and TUs connect to serial ports with little difficulty. Many also work with TTL interfaces. Check the MCP or TU specifications *before* you make your purchase. Make sure it will communicate with your computer!

Software is another area of concern. To talk to your MCP or TU you need a *terminal* program. Why is it called a terminal program? Well, when your computer is communicating with the MCP or TU, it's functioning as a data terminal. It displays information received from the microprocessor and allows you to respond via the keyboard.

Large business-system computers often work in the same manner. The main processor is tucked away in a room by itself while the employees communicate with it using data terminals scattered throughout the building. There's a microprocessor inside your MCP or TU and it also needs a terminal to communicate with the outside world. Your computer can't provide that function by itself—unless it receives the proper instructions. In other words, you can make it *behave like a data terminal* by running terminal software.

Terminal software doesn't have to be complex or expensive. If you're presently using a program to access telephone bulletin board systems or databases (such as CompuServe or Prodigy), you can use the same software to talk to your MCP or TU. There are also RTTY/AMTOR-specific programs that provide many useful features—including "help" messages, split-screen displays for incoming and outgoing text and so on. A number of software sources are listed in the RTTY/AMTOR Resource Guide. I encourage you to write for their catalogs and see what's available for your computer. I think you'll be pleasantly surprised to discover how clever—and inexpensive—many of the programs are.

But I Don't Own a Computer!

Don't despair if you're "computationally impaired." The

market is glutted with new and used computers in every price range. Who knows? Maybe you can get by *without owning a computer at all.* (Say *what*? Read on and you'll see!)

So what kind of computer do you need? The answer is up to you. IBM PCs and compatibles are the standard of the Amateur Radio community, but if you choose an Apple, Commodore, Atari or other machine, you won't be left out. The greatest incentive for owning a PC is the huge amount of Amateur Radio software that's available.

If you have the funds to buy new equipment, buy the best computer you can afford. Believe me, you won't regret it. Try to look beyond Amateur Radio applications. Would you like to do your household finances on the computer? Maybe you'd like to play some games, too. Buy a machine that will suit your needs for years to come. Here's a quick shopper's checklist for new computers:

❏ How much software is available for the computer? It may be the greatest machine ever built, but it won't be able to do much without software! Keep ham applications in mind. Can it run the programs that appeal to you the most?

❏ What kind of monitor is available? CGA used to be the popular choice, but it's been superseded by high-resolution VGA and Super VGA monitors. When it comes to RTTY and AMTOR, the distinctions aren't very important. In fact, CGA is fine. However, if you're thinking of running games and other graphics-oriented software, buy the VGA or Super VGA option.

❏ Check out the disk drives. Does it have at least two floppy drives? Good. How about a hard disk? Any new computer worth owning should have a hard drive—and the bigger the better.

❏ How about memory? 640 KBytes of RAM (Random

Access Memory) is the practical minimum, although more is always better!

❑ How fast is the computer? The speed of a computer is expressed by the speed of its clock, in MHz. The faster the clock, the more efficient the computer. For RTTY/AMTOR applications, speed isn't all that important. (I use an old 7-MHz IBM-XT for my RTTY and AMTOR operating. It's fast enough for me.) Once again, however, you have to consider the future. Complex software runs best at higher speeds. Many types of software now *require* high-speed machines to perform properly.

The guidelines we've just discussed concern *new* computers, but there are an awful lot of older machines waiting to be purchased. Visit a hamfest or computer flea market and you'll find many used computers for sale. Refer to the new-computer guidelines while you're bargain hunting. If you're willing to compromise on features (128 K of memory instead of 640 K, one disk drive, etc), you can pick up a used Tandy Color Computer, Commodore, Atari and others for well under $100. Raise your sights a bit higher and you'll discover Apples and IBMs for less than $300, depending on the machine. Remember the standard warning when shopping at flea markets: let the buyer beware! Plug it in and make sure it works *before* you part with your cash.

RTTY/AMTOR Without a Computer

Yes, it *is* possible to operate RTTY and AMTOR without the services of a computer. When you use a computer with terminal software, you're telling the computer to behave like a data terminal. Well, why not use the real thing?

At hamfest and computer flea markets you'll often find used data terminals for sale. Many amateurs take this approach to RTTY/AMTOR operating because it's very inexpensive. At one hamfest, I saw a fellow selling a truckload of used terminals for $10 each!

As you've probably guessed, there are some potential pitfalls in the used-data-terminal game. Consider the following factors carefully before you reach for your wallet:

❏ Check the condition of the terminal. Look at the screen closely. Do you see shadowy lines where the phosphor coating has deteriorated after years of constant use? (On some overused terminal screens you'll actually be able to see horizontal lines where the text was displayed!) This is a sure indicator of a terminal that's seen better days.

❏ If you want to save the text from your RTTY/AMTOR conversations, a data terminal may not be the best choice. Most terminals do not contain disk drives. Some terminals *do* support printers. You may have to pay a little more, but at least you'll have the means to save text.

❏ When you purchase a data terminal, you get the terminal software that's stored in its permanent ROM (Read-Only Memory). If you want to try other RTTY/AMTOR programs, you're out of luck. Remember: a data terminal is *not* a computer.

Used data terminals are excellent if you want to get started in RTTY/AMTOR while keeping your costs to a minimum. If you don't think you'll ever have a need to save data for later use, a terminal is a fine alternative to a full-fledged computer. This is especially true if you have no other use for a computer. On the other hand, you'll be missing out on much of the fun and versatility of the HF digital modes by not having a computer in the shack.

You *Can* Get There from Here: MCPs and TUs

Computers (or data terminals) and SSB transceivers are about as incompatible as you can imagine. Send digital data directly to a transceiver and you'll get the radio equivalent of

"Huh?" ("Arrgh" may be more like it!) Send receive audio from a transceiver directly to a computer and you'll get a similar response. The transceiver is analog and the computer is digital. How shall the twain meet?

Some computers incorporate analog to digital converters—and vice versa—and they can be used to send and receive RTTY on a limited basis. (The Tandy Color Computer is a typical example.) However, most computers need an external device to function as the bridge between the analog and digital worlds. I introduced you to multimode communications processors (MCPs) and terminal units (TUs) in Chapter 1. Now it's time for another look.

MCP or TU—What's the Difference?

Multimode communications processors are byproducts of the packet radio revolution. As packet terminal node controllers (TNCs) became popular, the market was ripe for a device that included packet, RTTY, CW, AMTOR and other modes in a single box. The manufacturers responded and before long the first MCPs made their appearance.

Since the mid '80s, multimode communications processors have gained a strong foothold in the amateur community. Their main advantage is convenience. By sending commands from a computer or data terminal, the operator can jump from one mode to another in a split second. Several MCPs feature special companion software that heightens the convenience factor even further. For example, when using Kantronics' *Hostmaster* software with the Kantronics *KAM*, it's possible to operate packet while *simultaneously* operating RTTY, AMTOR or CW.

The digital signal processing revolution has had an impact on multimode communications processors as well. DSP-based MCPs are truly multimode devices! Rather than using specific internal hardware (various integrated circuits)

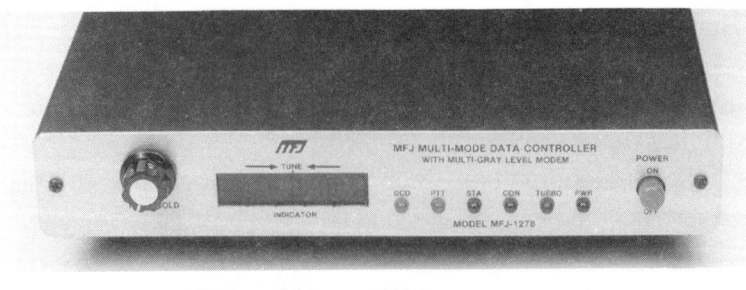

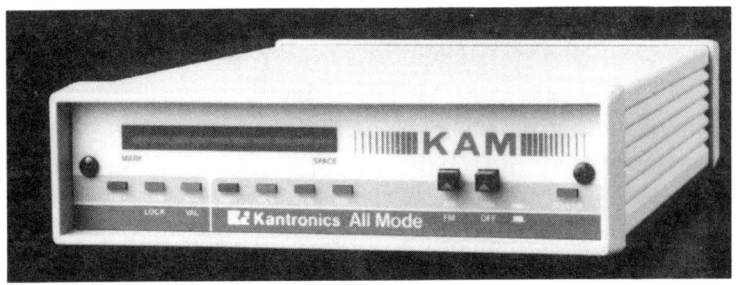

Multimode communications processors (MCPs) such as these offer RTTY, AMTOR, packet, CW, NAVTEX and other modes in a single device.

to process signals, DSP units use *software* to accomplish the same thing. Theoretically, a DSP multimode communications processor will never be obsolete—regardless of new modes that may appear in the future. To operate in a new or different mode, all the DSP device needs is new software!

How does a terminal unit (TU) differ from an MCP?

Terminal units—also called RTTY modems—are dedicated strictly to the task of converting audio into data and vice versa. Most terminal units are RTTY-only devices; they do not operate under the required protocols for AMTOR or other digital modes.

Terminal units have been at the heart of RTTY stations for decades. They've evolved over the years, incorporating the latest electronic technology into their designs. While MCPs offer good RTTY performance, terminal units are *optimized* for RTTY. TUs maintain reliable communications in conditions where MCPs would be hard-pressed to deliver even partial text!

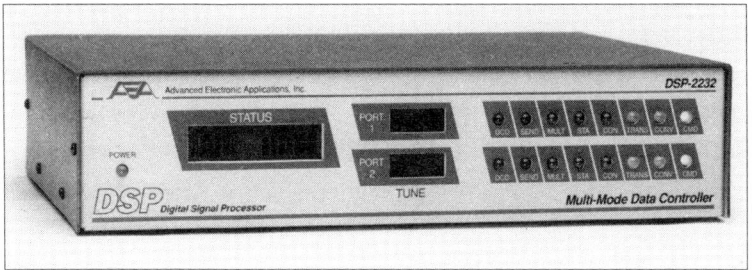

AEA's DSP-2232 is typical of the new series of multimode communications processors that incorporate digital signal processing (DSP). All of the data processing is accomplished with *software*, not hardware. As a result, adding a new mode is as simple as adding new software!

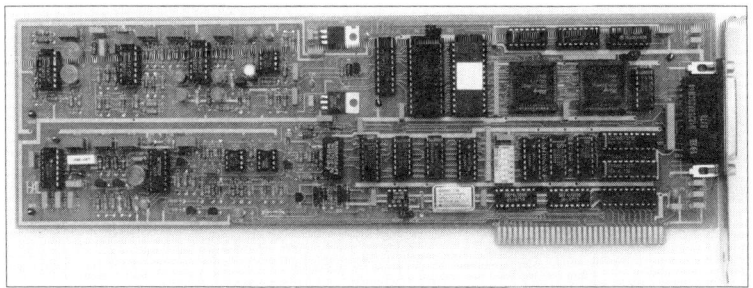

The HAL Communications PCI-3000 is a RTTY/AMTOR controller that fits *inside* an IBM PC or compatible.

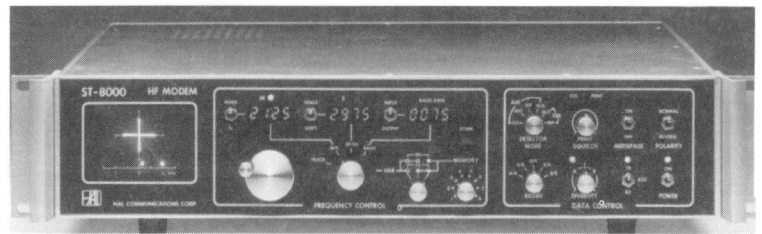

The HAL Communications ST-8000 is a RTTY modem for the serious RTTY operator. Its circuitry is designed to offer exceptional performance even under marginal conditions.

RTTY operators who are serious DX hunters, contesters or traffic handlers favor terminal units for their exceptional performance. TUs can also be used for AMTOR by installing special converters.

Radios and Amplifiers

When it comes to RTTY, just about any SSB transceiver will do the trick, regardless of age. AMTOR ARQ, on the other hand, is less flexible because of its rapid transmit/receive switching cycles.

As a rule of thumb, HF transceivers manufactured after 1984 should be able to operate AMTOR reliably. This is not to say that AMTOR is impossible for older radios. I presently use a vintage 1977 Kenwood TS-820S transceiver on AMTOR with excellent results. Even my ancient Drake TR-4 transceiver can clatter its relays fast enough to make an AMTOR contact!

Some hams worry about relay failure when running AMTOR in older equipment. There is some cause for concern, but I wouldn't lose sleep over it. Relays are pretty hardy devices. When you're operating AMTOR, they may sound as though they're about to undergo what the military calls "energetic disassembly." Try to ignore it and watch your screen instead!

FSK vs AFSK

You can operate RTTY and AMTOR in the AFSK mode by feeding the audio output from your MCP or terminal unit directly to the microphone jack, or auxiliary audio input, of your SSB transceiver. Just be careful not to overdrive your rig. If your radio has a MIC GAIN control and an ALC meter, adjust the control to keep the tones from exceeding your maximum ALC level. *Do not* use speech processing or compression when operating AFSK RTTY or AMTOR.

What if your radio includes an FSK mode? Is it "true" FSK as we discussed in Chapter 1? In most cases, the digital pulses from your MCP or TU are used to key an audio oscillator/modulator inside the transceiver. What is labeled as "FSK" is often AFSK in disguise! Even so, the oscillator/modulator in the transceiver may be superior to the one in your MCP or TU. By using the FSK input, you'll gain a superior signal. The transceiver will also add sharper receive filters in the FSK mode, which is a big plus when operating on a crowded band (more about filters later).

Receive Audio

Receive audio is easy to obtain if the rig has an external speaker jack. You'll really be in HF digital heaven if you own a radio that features an auxiliary fixed-level output such as a phone-patch jack! It puts you in the enviable position of being able to supply a constant audio level to your MCP or TU regardless of your front panel VOLUME control setting. Once you've established contact, you can turn down the volume and continue without disturbing anyone else in the house. If your radio lacks an auxiliary audio output or external speaker jack, you'll need to tap the audio at the speaker. A simple **Y** connector will do the job nicely.

Are you Stable?

Operator stability is a matter for mental health professionals and will not be discussed here. When it comes to radios, though, *frequency* stability is a prime concern. RTTY is a very forgiving mode. Your radio can drift off frequency and you'll still be able to copy clean text—to a point. When you've drifted too far, a gentle nudge of the VFO control will put you back on target.

Don't expect the same flexibility with AMTOR. If your radio drifts too far off the frequency, the link will cease altogether! Quick action will place you back on the proper frequency, but it may be too late.

Modern rigs employ digital synthesis and/or phase-locked loops to provide rock-solid stability. Older radios—particularly tube-type rigs—are not as stable and can drift quite a bit as they heat up. If you're using a vintage transceiver, allow about 15 minutes of warm-up prior to operating RTTY; 30 minutes for AMTOR.

As long as we're on the subject of frequency, I highly recommend transceivers with digital frequency displays. Why? If you're trying to communicate with a RTTY or AMTOR station on a particular frequency, it helps to have a display that's accurate and easy to read. For example, when I want to access the WA1URA/9 APLink system on the 20-meter band, I just dial up 14.071.30 MHz. There is very little guesswork. Of course, digital frequency displays don't always tell the truth (see the sidebar, "Frequency Displays Never Lie—Do They?").

Something's Burning

Output power is a major consideration with both RTTY

> **Frequency Displays Never Lie—Do They?**
>
> The amateur standard for specifying a RTTY frequency is to specify the frequency of the MARK signal. While this is a logical approach, the MARK frequency can be *different* from what your rig's fancy multidigit display tells you it is!
>
> If you're using your transceiver in the LSB mode, your digital display indicates the *suppressed carrier frequency*. In most cases you can subtract 2125 kHz to determine your MARK frequency. On the other hand, if you're in the FSK mode, you'll discover that calculating the exact frequency is *not* a matter of simple subtraction! It all depends on what rig you're using.
>
> Some radios (ICOM and Ten-Tec units, for example) show the MARK frequency. Others indicate the SPACE frequency (TS-930 and TS-940 in particular). Others show the *suppressed carrier frequency* (just like LSB operation). And still others show F_0—the imaginary center frequency between MARK and SPACE. (MARS stations specify F_0.) If in doubt, read your manuals.
>
> It's also important to note that digital frequency displays are *not* frequency meters! Usually, three or four oscillator stages, in addition to the VFO, determine the rig's output frequency. If a frequency error occurs anywhere other than the VFO, it may *not* be evident in the display. Your display reading can easily be several hundred or even a few thousand hertz off! Buying the "high stability option" (if available) will improve the frequency stability of your transceiver, but it usually won't correct your display calibration. If you really want to know your exact RTTY frequency, buy a frequency counter, attach a short wire antenna and measure your MARK signal frequency while sending continuous MARK pulses. This technique works on *all* radios.

and Mode B (FEC) AMTOR. RTTY and Mode B AMTOR are *100% duty cycle* modes—meaning that the transmitter is keyed *continuously* during each transmission. (By contrast, CW typically has a 50% duty cycle and SSB even less.)

Making a long RTTY or Mode B AMTOR transmission is the equivalent of keying your transceiver at full output and holding it there for several minutes!

Some transceivers are designed to cope with this kind of punishment. They feature heavy-duty power supplies, cooling fans and so on. Make *very* sure your rig is rated for full output at 100% continuous duty—don't just assume it is! If it isn't, plan on operating with your output power reduced at least 50%. On the receiving end, the difference between full output and 50% output is insignificant. The penalty for pushing your radio beyond its limits is severe. We're talking expensive tube or transistor replacements!

Turning up the Power

Most hams enjoy a lifetime of RTTY and AMTOR operating using only the output power their transceivers provide. In fact, when propagation conditions are good and

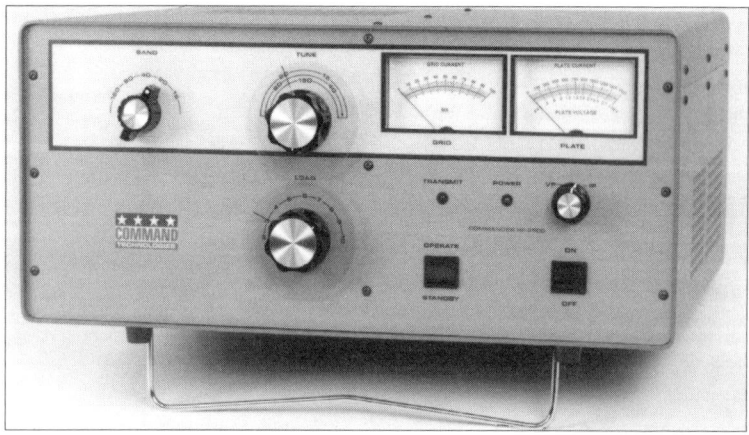

If you want to run high-power RTTY or AMTOR, you'd better buy an amplifier that can withstand 100% duty cycle transmissions. High transmit/receive switching speeds are also required. This Command Technologies HF-2500 amplifier is typical of units rated for heavy duty RTTY/AMTOR service.

interference is minimal, AMTOR is an exceptional performer at low power levels. (I once made an AMTOR contact with just a couple of watts to a dipole antenna!) But what if you're gripped with the urge to burn the airwaves with raw, unadulterated power? You need an *amplifier*!

Choosing an amplifier for RTTY and AMTOR is not a matter to be taken lightly. Remember the 100% duty cycle problem? Well, your amplifier must also be rated for 100% duty cycle operation if you intend to run it at full output on RTTY or Mode B AMTOR. If you think repairing a transceiver is expensive, wait until you cook a couple of amplifier tubes or fry a high-voltage power supply.

Don't forget about switching speeds. The amplifier must be able to switch from transmit to idle very rapidly (in a few *milliseconds*) if you expect to use it with AMTOR ARQ. Not all amplifiers are able to accomplish this, so do your research carefully. When shopping for an amp, look for "full QSK" capability. That usually—but not always—means that the unit can switch fast enough for AMTOR.

I won't deny that there are advantages to running high power with RTTY and AMTOR. In contests and DX competitions, it sometimes takes that extra push to cut through the interference. High power can also make the difference when propagation conditions are poor. Still, your amplifier may be more of a headache than a help. Producing more RF can cause interference to your home electronics equipment (TVs, telephones, stereos, VCRs) *and* your neighbor's. That extra RF may also find its way into your computer, MCP or TU. If you're operating AFSK and overdriving your transceiver, the amplifier will make the problem even worse—and may attract the attention of the FCC!

So am I steadfastly against high-power RTTY and AMTOR? Not at all. Just make sure you consider every aspect

Building Your RTTY/AMTOR Station   2-15

before you spend your money, or reach for the ON switch.

Filters

Take a moment to glance at the RTTY/AMTOR frequency guide in Chapter 3 (Table 3-1). You'll see right away that RTTY and AMTOR are squeezed into fairly tight subbands. If you add increasing PacTOR and CLOVER activity, the playing field can get crowded in a hurry! This is particularly true on 20 meters, the most popular RTTY/AMTOR band.

So how do you separate the signals you want from the signals you don't? You need a device that rejects as much interference as possible without mangling the signal you want to receive. If you're lucky, your HF transceiver will be equipped with a 500-Hz *IF filter* that you can select from either the LSB or FSK mode.

Many transceivers also offer selectable *audio filters*. If an audio filter is well designed, it will pass RTTY and AMTOR tones while reducing or rejecting signals from other stations operating near your frequency.

Some transceivers are not as flexible as others when it comes to filter selection. When operating AFSK RTTY, for example, most hams place their radios in the lower-sideband (LSB) mode. Many transceivers restrict you to a 2.4-kHz SSB filter in this mode. This is fine if the band isn't too crowded. Under congested conditions, however, a 2.4-kHz filter just doesn't do the job! This is another reason why many RTTY/AMTOR operators use the FSK mode, if it's available. In the FSK mode, the transceiver may provide a narrower IF filter.

If the IF filters in your rig are either too narrow or too wide, don't lose hope! You may be able to buy an IF filter from the manufacturer and install it in your radio. If that option isn't available, you'll have to consider an *outboard* audio filter.

Tunable outboard filters such as this Autek unit will greatly improve your ability to copy RTTY and AMTOR signals in crowded band conditions.

An outboard filter is easy to install and use. You simply feed the receive audio to its input jack; the filtered audio is available for your MCP or TU at its output jack. Until recently, all active audio filters were based on resistor/capacitor networks. By changing the value of one component or the other, you change the filter's bandwidth. As far as the audio signal is concerned, it's a bit like opening or closing a window depending on how much air you want in the room. These filters work very well for everything from SSB to RTTY/AMTOR to CW.

Digital Signal Processing, or *DSP*, has caused a revolution in audio filter design. DSP filters do not use resistor/capacitor networks. Instead, the incoming audio is converted into digital data for processing by specialized DSP software. The software searches through the signal data, rejecting noise and interference according to the desired bandwidth. The result is translated back into audio for use by your MCP or TU. DSP filters are exceptional when it comes to rejecting noise and certain types of interference. Ignition

Digital signal processing (DSP) has revolutionized audio filters. Here is a unit designed by Dave Hershberger, W9GR. It was featured in September 1992 *QST*. You can build it yourself for less than $150! All it takes is a flick of the switch to select optimal RTTY/AMTOR filtering.

noise from a nearby car, for example, is removed completely. Some DSP filters "seek and destroy" interfering carriers—such as when someone decides to tune their radio on or near your operating frequency!

You can make your own audio filters or buy them. There are a wide variety of designs to choose from. The most important thing to keep in mind, however, is that audio filters are not miracle devices. They'll make it much easier for you to operate on crowded bands, but they can't reject *all* interference. Even a DSP filter can't block a signal that's sitting right on your frequency!

Narrow filters (audio or IF) are *not* required to operate RTTY or AMTOR. I've managed to operate in contest conditions with wide SSB filters, although my score wasn't very impressive. The best advice is to get some on-the-air experience before you purchase an extra filter. Depending on where you operate and under what conditions, you may decide you don't need it after all.

Putting it all Together

As you begin installing your RTTY/AMTOR components, remember to establish good RF ground connections between each piece of equipment. RFI can be a major headache with solid-state devices. Computer birdies, for example, can make reception miserable. Your transmitter can also wreak havoc with your computer. I've used 1/2-inch copper braid for my ground connections with good results. Many RFI experts now advocate using 1/2-inch wide copper *straps*. Regardless of the material you choose, make the ground connection your *first* priority.

Keep all cables as short as possible. A 20-foot audio cable makes a marvelous antenna for RFI. Try to locate all equipment close together and limit cables to no more than six feet in length. Inexpensive audio cables will work just fine. However, a piece of RG-58 coax with phono connectors has superior shielding. (Don't waste your money on expensive, gold-plated audio cables.) If you have an RFI problem, try improving the grounds and cable shielding first. Contrary to audio grounding techniques, ground the cable shields at *both ends*.

As you hook up your equipment, read your MCP/TU, computer and transceiver manuals thoroughly. You'd be surprised at how much grief you'll avoid with an hour or so of light reading!

Pay special attention to the following . . .

❏ If you need to make custom cables, make sure you have the correct wires connected to the correct pins. Faulty wiring is a major cause of start-up problems in RTTY/AMTOR stations.

A Handy Attenuator for AFSK Operating

The commercial standard for audio-frequency signals is a 0-dBm level (about 0.77-volts RMS) and 600-ohms impedance. Since most amateur equipment is designed primarily for voice transmissions, we rarely find a 600-ohm, 0-dBm audio input.

Microphone inputs usually require as little as 30- to 70-mV RMS (−20 to −30 dBm) for full RF output. This is 20- to 30-dB *below* the 0-dBm level provided by many MCPs and TUs. A schematic for a simple level-matching attenuator is shown in Fig 2-2. The attenuator provides a 20-dB signal reduction between the TU or MCP output and the microphone input. If you experience difficulty matching your TU or MCP output to your rig, this attenuator may provide the cure.

Make sure the attenuator is well shielded (an aluminum mini-box is ideal). Also, run shielded cables to and from the box. Place the attenuator close to the transmitter. Set the TU or MCP output level about 1/3 below its maximum setting and perform transmitter power adjustments using the rig's mike-gain control. If you experience RFI, try adding 0.001-µF disc-ceramic capacitors between the center pins of the attenuator connectors and ground.

C1 in the attenuator may not be required with some equipment. An increasing number of transmitters place dc on the audio input terminal for powering an electret-capacitor microphone, microphone preamplifier or other active devices. C1 blocks dc, but passes the audio signal.

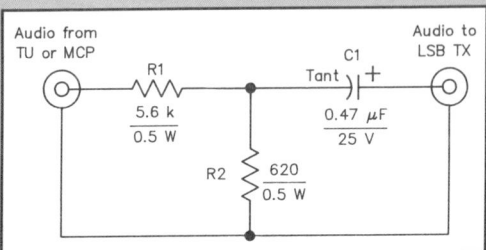

Fig 2-2—This easy-to-build audio attenuator is ideal for matching the output of your MCP or TU to your transceiver for AFSK RTTY or AMTOR. Chances are you won't need it but, as you can see, it's a simple project if you do!

Keying Older Rigs

MCPs and TUs use solid-state switching for transmitter control. Solid-state switching is fast and efficient. It's perfect for modern transceivers, but it can cause problems when used with older gear—especially tube-type radios.

Marrying today's technology to yesterday's equipment can be a challenge, but it's not impossible. One easy solution is to buy a small 12-volt relay and wire it as shown in Fig 2-3. The relay acts as an isolator between the MCP/TU and the rig. The MCP/TU keys the relay which, in turn, keys the transmitter. More elegant solutions are possible using solid-state devices. See "Cheap and Easy Control-Signal Level Converters" by James Galm, WB8WTS, in February 1990 QST, pages 24-27.

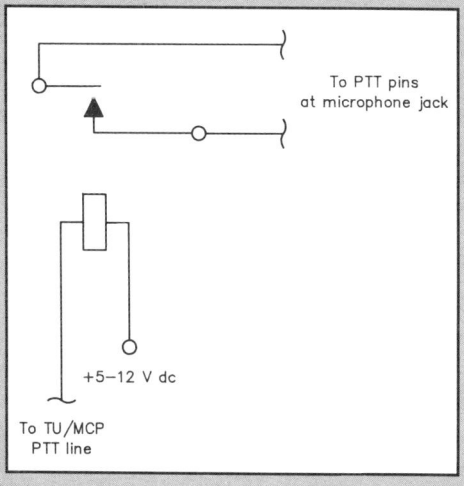

Fig 2-3—If you can't get your MCP or TU to key an older radio, an inexpensive relay will solve the problem.

❏ Your MCP/TU and your computer/terminal must communicate with each other at the same data rate. If you're using a computer with terminal software, you can change the software settings to match whatever your MCP or TU requires. Data terminals, however, often have a *fixed* data rate. In this case, you'll have to change the rate within the MCP or TU. Some MCPs use an *autobaud* routine that repeats a message several times at various data rates. When you see the message in plain text, you know the data rate is correct for your terminal or computer.

❏ Depending on the age of your radio, the keying circuitry inside your MCP or TU may not be compatible. You may wire everything properly, only to find that your transceiver will not switch to the transmit mode! Older radios often have problems with solid-state switching used by modern TUs and MCPs. See the sidebar, "Keying Older Rigs."

If you have your station up and running, there's nothing else to do but get on the air and start enjoying RTTY and AMTOR. The operating techniques for both modes differ considerably. Let's start with the easiest: RTTY.

CHAPTER 3
Your First RTTY Conversation

It's natural to feel tense during your first RTTY conversation. Communicating by keyboard may seem a bit strange unless you already have experience with packet radio or computer-aided CW. Even so, I think you'll be pleasantly surprised to discover just how easy RTTY can be!

Before attempting your first contact, spend some time copying RTTY signals. Take a look at the common RTTY/AMTOR subbands shown in Table 3-1. Depending on propagation conditions, you're likely to find plenty of RTTY

Table 3-1

RTTY/AMTOR Subbands

(MHz)
1.800 - 1.840
3.590 - 3.620
7.080 - 7.100
10.130 - 10.140
14.070 - 14.095   LSB
18.100 - 18.105
21.070 - 21.090
24.920 - 24.925
28.070 - 28.120

activity in the upper portions of these subbands. The 20-meter segment is especially popular. On weekends you'll encounter wall-to-wall signals here! Use your TU or MCP to eavesdrop on some of these conversations. In the process you'll learn your first RTTY/AMTOR skill: how to use your tuning indicator.

Am I Tuned In?

The answer to this question is easy. If you see meaningless gibberish—or nothing at all—you're not tuned in!

Every TU and MCP features some sort of tuning indicator, depending on the type of equipment you're using. In days gone by, RTTY operators would attach oscilloscopes to their terminal units and tune the signals until they saw the classic "crossed bananas" display (see Fig 3-1). As technology advanced, many terminal units included tiny built-in oscilloscopes that performed the same function.

MCPs often use LED indicators. The Kantronics KAM, for example, features an indicator comprised of several LEDs arranged in a horizontal bar. When the LEDs at opposite ends of the bar flash in sync with the RTTY signal, you know you have it tuned properly. Before you attempt your first contact,

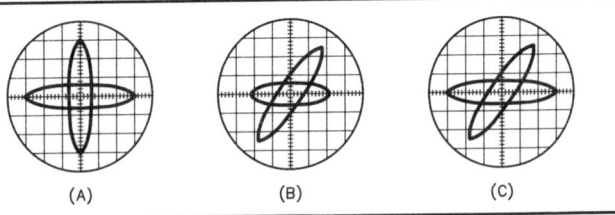

Fig 3-1—Oscilloscope-type tuning indicators produce patterns like these. Pattern A is the classic "crossed bananas," showing that the RTTY signal is tuned properly. At B the receiver is slightly off frequency, while C indicates that the transmitting station is using a shift that differs from the TU or MCP setting.

read the TU/MCP manual carefully to familiarize yourself with the operation of your indicator.

What Is That Signal?

As we discussed in Chapter 1, most RTTY operators use lower sideband transmissions with a 170-Hz frequency shift between the MARK and SPACE signals. The commonly used data rate is 60 words per minute, often expressed as 45 baud. But what if you stumble upon two operators who aren't conforming to "conventional" practices? Your indicator says you're tuned in properly, but nothing coherent prints on your screen!

It looks like you'll need to do a little detective work. Check the following:

❏ Is the signal "upside down"? The RF frequency of the MARK signal is usually higher than the RF frequency of the SPACE signal, but there is no law that dictates this standard. With most TUs and MCPs, all it takes is a push of a button or keyboard key to invert the normal MARK/SPACE frequency relationship.

❏ Are the operators really using 170-Hz shift at 45 baud? For example, many RTTY operators prefer to run at 75 baud (100 WPM) when exchanging lengthy files. To complicate matters, an operator may also decide to use an 850-Hz shift.

It may be of some comfort to know that these situations are uncommon. You may encounter RTTY at 75 baud or higher, but most operators stick to 45 baud. The use of inverted MARK/SPACE signals and odd frequency shifts is relatively rare.

Listen to the Action

Let's say you've successfully tuned in a RTTY signal. Unless you happen to be monitoring during a contest—which we'll discuss in Chapter 6—a typical RTTY conversation may look like this:

I FINALLY TOOK DOWN MY OLD VERTICAL AND REPLACED IT WITH A DIPOLE CUT FOR 40 METERS. YOU KNOW WHAT I FOUND OUT? I CAN FORCE FEED THAT DIPOLE WITH MY ANTENNA TUNER AND USE IT ON 20, 15 AND 10 METERS. N9GOR DE WA1MBK K

WA1MBK DE N9GOR....REALLY? THAT SOUNDS PRETTY INCREDIBLE. I HAD NO IDEA YOU COULD DO THAT. I BET THE SWR IS VERY HIGH ON THE OTHER BANDS SINCE THE DIPOLE WOULD NOT BE RESONANT ON THOSE FREQUENCIES. WHAT ABOUT THE RF YOU LOSE IN YOUR FEED LINE? WA1MBK DE N9GOR K

N9GOR DE WA1MBK...FEED LINE LOSS IS NOT A PROBLEM EVEN WITH A HIGH SWR. AS LONG AS YOU USE LOW LOSS COAX OR OPEN WIRE FEED LINE, THE LOSS IS SO SMALL IT DOES NOT MATTER. A GOOD ANTENNA TUNER IS THE KEY...N9GOR DE WA1MBK K

And so it goes, back and forth at a leisurely pace. When an operator is not sending data, you'll hear a continuous tone or a rhythmic *bee-bee-bee-bee* signal. The rhythmic signal is known as the *diddle*. The MCP or TU is simply switching back and forth between MARK and SPACE signals while awaiting more data from the operator.

The call signs at the beginning and end of each transmission are optional, as long as you identify at least once every 10 minutes. Still, some habits are hard to break and you'll find that RTTY operators often open and close their transmissions with an exchange of call signs. This is made easier by the fact that many modern MCPs and TUs feature automatic call sign exchange capability. You simply enter the

The impressive RTTY station of KC1YZ. Note the AEA model PK-232 MCP immediately in front of his keyboard (just to the left of his transceiver).

call sign of the other station *once*. After that, you can generate the entire <his call sign>-DE-<your call sign> sequence by pressing a single key.

Did you notice the **K** used to signify the end of each transmission? It's a RTTY custom to use CW prosigns in conversations. Depending on the operator, he or she may send a **K** (over to you), **AR K** (end of message, over to you) or **KN** (over to you *only*). When the conversation is over, it's common to use **SK** to signal the end of the contact. RTTY operators also adopt the CW custom of abbreviating words. In the sample conversation shown above, "frequencies" may be sent as "freqs."

Time to Pounce!

Enough monitoring! Now it's time for action. What if

you're eavesdropping on a QSO and you see that it's about to end?

N9GOR DE WB8IMY . . . I HAVE TO RUN, WAYNE. DINNER WILL BE READY IN AN HOUR AND I STILL HAVE NOT FINISHED MOWING THE LAWN. HAVE A GOOD WEEKEND AND I HOPE TO CHAT WITH YOU AGAIN . . . . N9GOR DE WB8IMY SK

WB8IMY DE N9GOR . . . NO PROBLEM. GLAD I HAD THE CHANCE TO DISCUSS ANTENNAS WITH YOU. I LEARN SOMETHING EVERY DAY. SEE YOU LATER . . . WB8IMY DE N9GOR SK

Looks like N9GOR is still available. Why not give him a call? Since you were able to copy the transmissions, chances are good that your MCP/TU is already set for the proper data rate, shift and MARK/SPACE frequency relationship. (A quick check never hurts, though!) Switch your MCP or TU to the transmit mode and start typing...

N9GOR N9GOR DE N6ATQ N6ATQ N6ATQ

N9GOR N9GOR DE N6ATQ N6ATQ N6ATQ K K

[switch back to receive]

Make sure to repeat your call sign several times. After all, N9GOR knows his own call, but he doesn't know *yours*! Send your transmission in several short lines rather than one long line.

If the other operator is able to copy your signal, you may see a response like this:

N6ATQ N6ATQ DE N9GOR N9GOR . . .THANKS FOR THE CALL. NAME HERE IS WAYNE WAYNE AND I AM LOCATED IN MILWAUKEE MILWAUKEE WISCONSIN WISCONSIN. YOUR RST RST IS 579 579 . . . BACK TO YOU . . . N6ATQ DE N9GOR K K

Split-Screen Software Makes it Easier

There is one feature that I consider to be almost indispensable to RTTY and AMTOR operators: split screen capability with a type-ahead buffer. As fancy as it sounds, this simply means that you'll have the ability to start entering your response while the other station is still transmitting.

In most cases, the incoming text from the other station is displayed in a separate area of the screen. Your responses can be typed into another area as you read the operator's comments. This may be confusing at first, but with a little practice you'll master the split-screen technique.

By using your type-ahead buffer, you'll be able to make comments or ask questions right away. If the other operator makes an interesting remark about his or her job, for example, will you remember to ask about it when it's your turn to respond? With type-ahead capability, you can ask your question the moment it occurs to you. When it's time for you to transmit, you can be sure it will be sent along with your other text.

The ability to type ahead also makes the conversation flow smoothly. When the other station has finished sending, all you have to do is switch to the transmit mode and continue typing. Your MCP or TU will start transmitting everything you've entered so far. If you're a good typist, you'll finish before your system "catches up" to you. As far as the receiving station is concerned, he or she sees nothing but smooth-flowing text—just like reading a commercial teletype!

I'm a terrible typist and type-ahead capability allows me to hide this fact from other hams. I keep my transmissions short so that I always finish my comments before the type-ahead buffer is empty. The other operators think I'm a great typist, of course! When I get too long-winded, however, my MCP sends all my pre-typed text before I can finish. The smooth-flowing transmission comes to a grinding halt and I'm exposed as the hunt-and-peck operator I really am!

Notice how N9GOR sends the important information *twice*. He doesn't know how well you are receiving his signal, so he wants to make sure that you won't miss his name, city, state and so on. This is always a good technique to use when you aren't sure of the signal path between your station and another.

Wayne has sent your signal report (RST) and it's 579. The first digit from the left is your readability (R), the second is relative strength (S) and the third is tonal quality (T). In this case, a 579 RST means that he is receiving you very well, your signal strength is moderate and your RTTY tones are good. A perfect RST would be a 599, but a 579 is fine. You can be reasonably certain that he is receiving everything you're sending. Go ahead and tell him who you are and where you are. Don't forget to give him a signal report, too!

[switch to transmit]

N9GOR DE N6ATQ . . . . HELLO WAYNE. NAME HERE IS CRAIG CRAIG AND I AM IN ESCONDIDO ESCONDIDO CALIFORNIA CALIFORNIA. YOUR RST RST IS 599 599. THIS IS MY FIRST RTTY CONTACT. I AM USING A MODEL PK232 MCP AND A KENWOOD TS-820S TRANSCEIVER. ANTENNA IS A DIPOLE UP 30 FEET. SO HOW COPY? N9GOR DE N6ATQ K

[switch to receive]

These are the preliminaries of most RTTY/AMTOR contacts. Some hams are very proud of their station equipment and will give you a brief rundown of their entire setup. In the earlier days of RTTY, messages such as these were created in advance and stored on reels of paper tape. When fed to a teleprinter, the holes punched in the paper tape were translated into MARK and SPACE signals for transmission.

If you own a computer, you can create and store

"canned" messages of your own and save them on diskette or magnetic tape. Depending on the type of software you're using, a single keystroke will send the entire message automatically! Many RTTY operators store and send their station descriptions in this manner, although the process is still known by its old namesake: the *brag tape*! Consult your software manual to learn how to create your own brag tapes and other stored messages.

Once you're past the introductions, the real conversation begins. If you're nervous and can't think of anything to say, start asking questions. What does the operator do for a living? What does he think about his equipment? How many DX contacts has he made? People enjoy talking about their interests and one question usually leads to another. Your conversation doesn't have to be technical; talk about anything that enters your mind!

Is There Anybody Out There?

If you can't find someone to talk to, consider calling CQ. You never know what you'll turn up!

[switch transmitter on]

CQ CQ CQ CQ CQ CQ CQ DE WB8IMY WB8IMY WB8IMY

CQ CQ CQ CQ CQ CQ CQ DE WB8IMY WB8IMY WB8IMY

CQ CQ CQ CQ CQ CQ CQ DE WB8IMY WB8IMY WB8IMY K K

[switch to receive]

A CQ should be long enough to attract attention, but short enough to avoid boring the other station. Repeat your call sign often so the operator on the other end has a decent chance of getting it right. You may have to send your CQ

> **Bad Habits**
>
> As you scan the RTTY subbands, you'll find some operators sending long streams of RYs at the beginning of their transmissions.
>
> **RYRYRYRYRYRYRYRYRYRYRYRYRYRYRYRY CQ CQ CQ CQ CQ CQ CQ CQ DE WB8QVC WB8QVC WB8QVC K K**
>
> This is another artifact from the early days of RTTY when it was necessary to make sure that mechanical teleprinters were ready to copy a transmission. In the modern era of computers and data terminals, this is unnecessary. Many operators find these long RY streams highly irritating since they do nothing but waste time and band space. Try to avoid the RY habit. If you have something to say, such as calling CQ, go ahead and say it. You don't need to send a meaningless string of letters in advance.
>
> You may also find operators who make frequent use of the date/time function included in their MCPs or TUs. Many units have this feature but, thankfully, most operators don't use it except in contest or public service situations. What does it do? It inserts the time and date (local or UTC) as part of the transmitted text. Usually it appears at the end of a transmission, but may pop up at the beginning. You may also see it at the end of stored messages.
>
> Unless you're in desperate need of a clock, do you really want to know what time it is? Probably not. Now turn the tables. Do you think the other station cares to be informed of the date and time? I think you get the idea! Time and date "stamping" is a waste of RF in most cases. Veteran RTTY operators consider it a nuisance, although most will be too polite to tell you!

more than once before you're noticed. Storing your CQ on disk makes it easy to send it again without retyping.

If you don't receive an answer, don't lose hope. Just move to another frequency or band. Before calling CQ, it's good practice to ask if the frequency is in use. Perhaps you've just tuned onto a frequency and it *seems* to be unoccupied. Don't let appearances fool you! You often hear only one side of a conversation—and that side may be listening at the moment!

Checking the frequency is easy. For example, I'd send:

QRL? QRL? DE WB8IMY WB8IMY K K

If the frequency is occupied, someone will let me know right away. On the other hand, if no one replies, I can assume it's safe for me to go ahead and call CQ.

It's a DX Pileup!

Sooner or later you're bound to encounter the fascinating phenomenon known as the *DX pileup*. You'll know when you've found a pileup because it will sound like pure pandemonium!

Pileups are the result of a desirable DX station coming on the air. What makes a DX station worthy of a pileup? If the operator is in a country that is not heard on the air often, that country is considered "rare DX." Any transmission from that part of the globe is a major event and it quickly attracts hordes of hams eager to make a contact.

The first CQs from a rare DX station snag the few hams who are lucky enough to be near the frequency at the time. More join the fray as they discover what's going on. DX *PacketClusters* also sound the alarm on the VHF frequencies and bring operators by the dozens. Within minutes you have a huge number of stations chasing the same goal: the hapless DX operator! As soon as he finishes a contact, everyone starts calling at once! You can imagine how this would sound in your receiver.

When you hear a pileup in progress, the first thing to do is monitor the exchanges. Determine the DX station's call sign and see if you can copy his signal. When the FR5ZU/G group began calling for contacts from Reunion Island in September of 1992, here is how it looked at my station.

DE XE1/JA1QXY ... XE1/JA1QXY ... XE1/JA1QXY ... BK BK

XE1/JA1QXY DE FR5ZU/G ... UR 599 599 BK TO U ... KN

QSL UR 559-559 TKS QSO DE XE1/JA1QXY

XE1/JA1QXY sends his call sign several times and is heard by FR5ZU. Signal reports are exchanged quickly and the contact is over in a matter of seconds! Notice the heavy use of Q signals and abbreviated words to speed the process. The DX station is fair game once again and two operators slug it out for the prize ...

DE W2JGR DE W2JGR DE W2JGR DE W2JGR K

FR5ZU/G DE NJØM NJØM NJØM PSE K

NJØM uses the traditional approach while W2JGR tries repeating his call sign preceded by **DE** ("from"). His tactic pays off and he wins.

W2JGR DE FR5ZU/G ... GOOD MORNING ... UR 569 569 ... BK

FR5ZU/G DE W2JGR ... TNX ... UR 579 579 NAME JULES ... QSL??? BK

FR5ZU passes along a signal report and shoots it right back to W2JGR. W2JGR gives his report and sends his name (Jules) as well. At the end of his transmission, he asks if the DX station copied everything ("QSL?").

QSL ES 73 ... FR5ZU/G QRZ KK

FR5ZU sends a quick "QSL" to mean, "Yes, I got it all" along with his best wishes. He immediately sends QRZ to signal that he is ready for another contact. That's NJØM's cue to try again!

DE NJØM NJØM NJØM PSE KK

What you've seen here is less than five minutes of a DX pileup that lasted over an hour! Pity the DX operator at the other end of this melee. He has to do his best to sort out readable call signs among the rampaging signals. Sometimes the interference is so severe, he sees nothing but a wild jumble of letters on his screen.

The best you can do is be patient and keep calling. The rules of DX courtesy say that you shouldn't transmit if a contact has already been established. If you keep transmitting, you may get the attention of the DX station in a way you'd never expect—he'll refuse to answer you for the remainder of the operation! Also, if the DX operator tries to control the mayhem through techniques such as working stations by call sign areas (1s, 2s, 3s and so on), don't buck the system. All you'll manage to do is anger the person you're trying to contact!

It's a "Split Decision"

What if you discover a rare DX station, but you don't hear a pileup? No matter how long you listen, you only seem to copy one side of the conversation—his! This is the telltale sign of a DX station that is *working split*. In other words, he is listening on one frequency and transmitting on another.

For some DX stations, working split is the only way to manage a pileup. This is especially true when a pileup gets too large and begins to disintegrate into chaos. Without the ability to work split frequencies, the DX station may be buried under a torrent of competing signals. Even if he manages to sort out a call sign, making contact is difficult because of interference from other stations who are continuing to call.

A good DX operator will always make it clear that he is working split and will indicate where he's listening for replies. You may see something like this:

CQ CQ DE 5U7M 5U7M UP 10

or . . .

QRZ DE 5U7M LISTENING 21.085 TO 21.090

The DX operator, 5U7M, is telling everyone that he is listening for calls 10 kHz above this frequency ("UP 10"), or that he is listening between 21.085 and 21.090 MHz specifically.

To contact a DX station using a split-frequency scheme, you'll need a transceiver that can transmit on a frequency other than the receive frequency. Many modern transceivers feature *dual VFOs* for this purpose. Other rigs have the capability to use a *remote VFO* (a second VFO in a separate enclosure).

If you have remote or dual-VFO capability, leave your

KB2BBW takes RTTY on the road! With his Toshiba laptop computer, AEA MCP and Uniden 10-meter transceiver, he has a fully equipped, 10-meter mobile RTTY station.

receiver tuned to the DX station and move your transmit frequency to the desired spot. Check your equipment carefully and make sure you know exactly where you are transmitting and receiving. Whatever you do, *don't transmit on the DX station's frequency.*

What's in the Mailbox?

Live conversations are the bread and butter of RTTY, but other activities may keep you away from the keyboard more often than you'd prefer. If a friend tries to contact you, he or she is out of luck, right? Not necessarily!

RTTY operators have enjoyed *autostart* capability, in one form or another, for many years. Autostart is simply the

---

Unattended Operation

If you're thinking of setting up a RTTY MSO of your own, make sure to consider the legal factors. At the time of this writing, it is illegal to operate a station on the HF bands without a control operator present at the location shown on the license. Assuming that you're the control operator, you don't have to be sitting in front of your radio at all times. However, you must be able to monitor the activity of your station and be able to shut it down immediately if necessary.

Some hams work out of their homes and this provides an ideal opportunity to operate a RTTY MSO in their shack. They're usually close enough to keep an eye on the activity and they can easily deactivate the system if something goes haywire. The same is true if you operate your station by remote control, either through a VHF, UHF or microwave link, or by a telephone connection.

On the other hand, being a control operator does not mean you can sleep on the job! If you want to hit the sack, or otherwise be unavailable, you'll need to shut down your MSO or enlist the aid of someone else who is capable of being a control operator.

ability of a RTTY station to operate in an automated manner without the direct intervention of a control operator. A station operating in autostart mode will activate only when it receives a special autostart call sign. Typical autostart call signs consist of up to 7 letters (no numbers). For example, the autostart call sign of WB8ISZ would be WBISZ, or whatever combination he decides to program into his MCP or TU.

The most basic autostart system is receive-only. If I wanted to send a message to WB8ISZ, I'd tune to the proper frequency and send:

WBISZ
WBISZ
WBISZ
WBISZ

Then, while still transmitting, I'd continue to send my message and end it with **NNNN** on a line by itself. If WB8ISZ receives the signal, my message will be waiting for him when he returns. Unfortunately, with this type of basic autostart system, I have no way to know if my message made it through. I just have to hope for the best!

Hams soon recognized that this situation wasn't very satisfactory. Those who had computer-programming skills began writing software that allowed response and interaction—not just passive reception. The RTTY Message Storage Operation—or *MSO*—was born!

When you send an autostart call sign to an MSO, the system responds and offers several commands you can use (see Table 3-2). Many RTTY MSO operators park their systems on the National Autostart Frequency, or *NAF* (14.087.7 MHz, suppressed carrier frequency, 14.085.6 MHz, MARK frequency).

On the National Autostart Frequency, the custom is to use MSO call signs composed of "MSO" followed by the last three letters of a call sign. For example, to contact the WB9GRP MSO, you'd send MSOGRP. Any other combination can be used, however. WR1B might use

Table 3-2

Common RTTY MSO Commands

Note: Command formats may differ from one system to another. Use these commands as a general guide only.

Command	Description
.OPEN	Opens mailbox function to begin message entry
.END	End message entry (send on separate line)
.MSG	View your message before saving it
.SAVE(filename)	Save your message to disk under the filename of your choice
.READ(#)	Read a message (# = message number)
.DEL(#)	Delete a message (# = message number)
.DIR	Show the list of available messages and other files
.110	Switch the MSO to 110 WPM ASCII
.100	Switch the MSO to 100 WPM (75 baud)
.60	Switch the MSO to 60 WPM (45 baud)
.EXIT	Deactivate MSO
.FIND(filename)	Download a file
.USERS	List a log of recent users
.SYSOP	Call the system operator to the keyboard

MSOWRB as his autostart call sign. Most of the MSOs on the National Autostart Frequency operate at 75 baud (100 WPM).

You'll find RTTY MSOs on other frequencies as well. Not all systems use the same software, so you may have to feel your way around the commands and autostart call sign formats. Let's say that you want to access the NF8F MSO. You'd send his call sign preceded by a period:

.NF8F
.NF8F
.NF8F
.NF8F

If you're successful, you'll be rewarded with something like this:

SYSTEM ACTIVATED - 08:21 PM UTC
PLEASE LOG ON NOW - USE: .LOGYOURCALL

(EXAMPLE) .LOGNF8F

NO SPACES BETWEEN .LOG AND YOUR CALL

DE NF8F MAILBOX
K-K-K

Looks like you'd better obey the instructions and log in!

[switch transmitter on]

.LOGWB8IMY

[switch back to receive]

Be patient. Some MSOs won't respond immediately. They need time to check disk files and prepare for a new user—you!

WB8IMY - HAS BEEN LOGGED
USER NUMBER: 2895

LAST USER: WKØF -
PLEASE STAND BY . . .

SCANNING DIR . . .

SORRY! NO MAIL FOR YOU TODAY

DE NF8F MAILBOX
K-K-K

Now you're logged into the MSO. It's your turn to make a request. Switch on your transmitter and send . . .

.DIR

. . .to request a directory of files you can download to your system. Notice that a period precedes the "dir" command. Switch back to receive and watch as the MSO responds to your request.

#——:FILES——
  Ø   SYSTEM INFO
  1   US TOWER

2   NEED IBM PROGRAM
3   NS8S DE NF8F
4   WANTED PK-232 OR EQUIVALENT
5   W9HSQ DE KC9PX
6   AIRLINE PILOT
7   WB8CKI
8   Q1PS
9   NAMES/CALLS (USERS)
10 WA0IFH DE PH
11 WA0IFH DE W8PH
12 KE9MG
13 WY4Z DE K9ODF
14 NEED C64 PROGRAM
15 KD9QM DE WD4MKQ
16 WD4MKQ DE NF8F
17 DSC CMD ON PK-232??
18 KO5T
19 PHONE PATCHES

DE NF8F MAILBOX
K-K-K

The MSO has turned it back to you after transmitting its list. Since this is your first contact with the MSO, you should read message #0 to learn more about the system. All you need to do is send the .READ command and the number of the file you want to see.

[switch transmitter on]

.READ0

[switch back to receive]

NF8F LOADING MSG -0-
WAIT PLEASE . . .

WELCOME TO NF8F MAILBOX..

A LITTLE INFO ON THE SYSTEM.

THE PROGRAM SUPPORTS 21-16 DISK DRIVES UP TO 160K MESSAGE STORAGE PER DISK DRIVE.THE SYSTEM RUNS ON A COMMODORE 64 AND A CP-1 TU ...

THE SYSTEM RUNS 40 WATTS INTO A 3 ELE TRI BANDER UP 50 FT.

THE SECOND DISK DRIVE WILL NOT ALLOW USER TO DELETE FILES. USE THIS DRIVE FOR PICTURES/BULLETINS/JOKES ETC ...

THE QTH IS DAYTON, OH

PLEASE DELETE YOUR FILES WHEN YOU HAVE READ THEM.

ALL IDEAS SUGGESTIONS WELCOME..

73 FROM SYSOP -:-VIC-:-

DE NF8F MAILBOX
K-K-K

Well, that's informative, isn't it? The NF8F MSO looks like a friendly system that you'll want to visit again. If you want to sign off for now, just use the .EXIT command.

[switch transmitter on]

.EXIT

[switch back to receive]

TNX FOR USING THE SYSTEM!

SYSTEM DEACTIVATED - 08:31 PM UTC

WB8IMY DE NF8F - TIME USED: 9 MINUTES

DE NF8F MAILBOX

DEFAULTING TO 60 W.P.M.

SEND .NF8F TO ACTIVATE MAILBOX

K-K-K

```
                            .                          .
                          )                         (
                        )                         (
                      )                         (
          (0000000           )             (         000000)
          (0000000000          )         (         0000000000)
          0000000000000          )     (         0000000000000
          (000        0000        )   (         0000        000)
          000  HHHHHHHH 0000        ) (          0000 HHHHHHHH 000
          000 HHHHHHHHH 0000         ) M (       0000 HHHHHHHHH 000
          000 HHH      HH 000        MMM        000 HH      HHH 000
          000 HH   000   HH  000    MMMMM    000  HH   000   HH 000
          000 HH  OHHHO  HH   000   MMMMM   000   HH  OHHHO  HH 000
          000 HH  OHHHHO HH   000   MMMMM   000   HH OHHHHO  HH 000
          (00 HH  OHHHHO HH   000  MMM 000  HH   OHHHHO  HH 00)
          000 HH  OHHHHO HH   000MMM000    HH  OHHHHO   HH 000
          (00  HH OHHCCCC HH  00MMM00   HH  CCCCHHO HH    00)
          000 HHH  OOCCCC    HH  OMMMO  HH   CCCCCOO HHH 000
          (00  HH   CCCCC    HH   MMM   HH   CCCCC    HH  00)
          000   HHHH CCCCCC  HHH MMM HHH  CCCCCC HHHH   000
             000 HHHH CCCCCC HH MMM HH CCCCCC HHHH 000
          000    HHHH        HHHMMMHHH        HHHH    000
          (00       HHHHHHHHHHHHHMMHHHHHHHHHHHHH       00)
          000       HHHHHHHHHH HHHMMMHHH HHHHHHHHHHH   000
          (00    HHHH         HHH MMM HHH        HHHH    00)
          000    HHH    CCCCC HHH MMM HHH CCCCC    HHH   000
          000  HHH  CCCCCCC   HH  MMM  HH   CCCCCCC  HHH 000
          (00   HH  CCCCCCC   HH  OMMMO HH  CCCCCCC  HH 00)
          000 HH   CCCCCCC    HH  OOMMOO HH   CCCCCCC  HH 000
          000 HH  CCCCCCC    HH  OOOMMOOO  HH  CCCCCCC HH 000
          000 HH  CCCCC   HH   000 MMM 000   HH   CCCCC HH 000
          000 HH  CCCCC  HH   000  MMM  000   HH  CCCCC HH 000
          000 HH  CCCC   HH  000     M     000  HH  CCCC HH 000
          000 HH  CC     HH  000           000  HH   CC  HH 000
          000  HHH      HHH 000             000  HHH      HHH 000
          000   HHHHHHHH   000               000   HHHHHHHH   000
          (00    HHHHHH   000                 000   HHHHHH    00)
           00             000                 000              00
           (00000000000)                        (00000000000)
            0000000000)                          (0000000000
             00000000                              00000000
```

                        BUTTERFLY

                  1979   RTTY   ART   CONTEST

         ORIGINATED BY SANDI CLARK, WB4VUE, HURT, VIRGINIA

Fig 3-2—RTTY artists use combinations of letters, numbers and punctuation to produce pictures such as the one you see here. These RTTY pictures are usually available for downloading from MSOs.

By using a RTTY MSO properly, you can read interesting messages from other amateurs, read bulletins on the latest events in Amateur Radio or leave messages for other amateurs. Many MSOs even have *picture* files available. These are images composed of letters, numbers or punctuation (see Fig 3-2).

CHAPTER 4

Time to Start Chirping with AMTOR

AMTOR is a one of the most fascinating HF digital modes. Its signals sound peculiar; even its name has a mysterious quality! (Doesn't the word "AMTOR" seem like something out of a science-fiction movie?)

In reality, AMTOR is a close cousin to RTTY. What sets it apart is its ability to detect errors. Unlike RTTY, AMTOR provides HF digital communications with a consistent level of accuracy.

Two types of AMTOR are used most often in Amateur Radio: *FEC* (or *Mode B*) and *ARQ* (or *Mode A*). When you hear those odd *chirp-chirp* signals, you're listening to ARQ AMTOR. FEC, on the other hand, sounds like very fast RTTY. Before you make your first AMTOR contact, let's take a moment to talk about both modes.

It Looks so Nice, I Sent it Twice

FEC stands for *Forward Error Correction*. This is a synchronized mode where every character is sent twice. The receiving station takes full responsibility for sorting out the errors. This means that an FEC transmission is strictly a one-way affair. The receiving station cannot ask for a repeat if it detects an error.

As the FEC transmission begins, the receiving station synchronizes itself to the transmitter. This synchronizing process may take as long as 10 seconds. When the first character is received, it's tested for the correct 4:3 bit ratio. (Remember our discussion of the bit-ratio test in Chapter 1?) If it passes the test, the character is assumed to be okay and is printed on the screen. If not, the system waits for the repeat and checks it, too. With any luck, the character will pass the test on the second try. Another failure, however, will cause the system to print a blank space or underline character.

Most AMTOR operators use FEC to send CQ in the hope of establishing an ARQ connection. Although it's not a common practice, you can also use FEC to carry on complete conversations. Like RTTY, *any* SSB transceiver is capable of operating in the FEC mode. Just be sure to watch your output power! FEC is a 100% duty-cycle mode. You may need to decrease your output by as much as 50% to avoid damaging your equipment.

Shaking Hands with ARQ

ARQ is a generic data term that means *A*utomatic *R*epeat re*Q*uest. In the ARQ mode, the *information sending station (ISS)* sends its text in bits and pieces. It transmits a group of three characters in a single burst and then switches to the receive mode. The *information-receiving station (IRS)* checks the characters for the 4:3 bit ratio and then transmits a control character. The control character means "Acknowledged. Send the next three" (*ACK*) or, "Not acknowledged. Repeat the last three" (*NAK*). If the ISS doesn't receive a reply (due to fading signals or interference), it repeats the characters anyway. Each station gets its turn to be the IRS or ISS.

It's fair to say that when you're using AMTOR ARQ there are really *two* conversations taking place. First, there is

Ed Joy, AA4TH, is justifiably proud of his high-performance RTTY/AMTOR equipment. Whenever Ed needs to run *power*, he gets the full legal limit from his modified Gates HFL-2500 amplifier (background)!

the conversation between you and the other amateur. Secondly, there is the conversation—often referred to as *handshaking*—going on between your hardware and the hardware at the other station. For as long as you and the other ham converse, your stations are locked into an intricate dance of signals. Of course, as in all dances, timing is everything!

ARQ Timing

AMTOR ARQ is precisely timed so that both stations know exactly when the other station is transmitting and

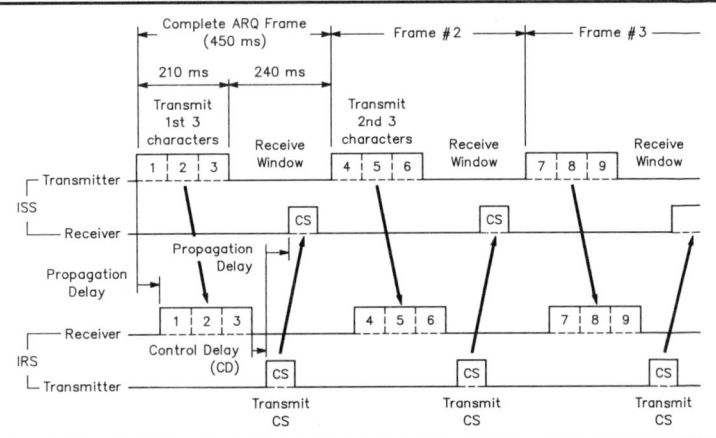

Fig 4-1—A typical AMTOR ARQ timing cycle. The dark arrows indicate the signal path from the information sending station (ISS) to the information receiving station (IRS) and vice versa. Notice how transmitted data does not reach the receiving stations instantaneously. This is caused by propagation delays. Propagation and equipment delays determine the maximum—and minimum—distance over which you can communicate.

listening. This is why high-speed transmit/receive switching is important when it comes to AMTOR equipment. A slow transceiver may still be in the act of switching when it should be listening or transmitting!

A typical ARQ transmission/acknowledgment sequence is illustrated in Fig 4-1. It may look complicated, but the concept is far simpler than the diagram!

A 450-ms ARQ cycle starts when the ARQ *link* is established. (Two stations enjoying an AMTOR ARQ conversation are said to be "linked.") The sending station (ISS) sends three AMTOR characters, which takes 210 ms. After receiving and checking the characters, the receiving station sends its ACK/NAK control signal (CS). The ISS has a 240-ms receive *window*. The CS character must arrive at the ISS before the window "closes," or the ARQ link will eventually fail.

The IRS requires 70 ms to send the CS character. That leaves 170 ms to spare (240 ms − 70 ms = 170 ms). Sounds like we've beaten the clock so far, doesn't it? The catch, however, is that we haven't considered all the possible delays that can take place. After all, even radio waves aren't instantaneous. They travel at the speed of light. The delay they induce is called the *propagation delay*. The propagation delay depends on the distance between the two stations.

Keep in mind that our goal is to communicate over varying distances. The distance may vary from "zero" (two stations a short distance apart) to halfway around the world. Taking the speed of radio waves into account, it takes about .067 seconds (67 ms) for a signal to travel halfway around the world. Since the ARQ signals must pass *both ways* within each cycle (ISS to IRS and IRS to ISS), we need *twice* that time for the total propagation delay (134 ms). If you subtract 134 ms from 170 ms, we still have 36 ms remaining. That's plenty of time—or is it?

Don't forget that equipment delays take a significant bite out of our spare time. The MCPs at each station add a *1 bit period* delay. That's 10 ms for each MCP, or 20 ms total. If slow-switching (60 ms) transceivers are used at both stations, an additional 120 ms is added to the delay (60 ms × 2 = 120 ms). Add the 20-ms MCP delays and our total equipment delay has now reached 140 ms! We've used all of our spare time with an extra 104 ms for good measure. (36 ms − 140 ms = −104 ms) This long-distance AMTOR link is doomed!

If both stations have very fast-switching transceivers (8 ms), the total equipment delay can be reduced to 36 ms (16 ms for the rigs and 20 ms for the MCPs). The signals will reach each station in time to maintain the link.

Now let's look at the opposite extreme. Consider two AMTOR stations side by side. The propagation delay is effectively zero. Let's also assume that the IRS sends its ACK/NAK control character immediately after receiving the ISS data. If the transmit/receive switching delays at the ISS

Flexible Timing

One possible solution to the AMTOR ARQ timing problem is the control delay (CD). This is the delay between the time the IRS receives a data pulse and the time it sends its ACK/NAK character. Here are some guidelines for adjusting your control delay:

❏ If you're working a nearby station (2500 miles or less), set your control delay in the range of 30 to 50 ms.

❏ If you're working a distant station (more than 2500 miles), set your delay to a lower value (10 to 20 ms).

❏ If you call a station and the station can't seem to maintain the link, try adjusting *your* control delay.

❏ If another station calls you and *you* can't maintain the link, the calling station must adjust his or her delay.

Not all MCPs will allow you to adjust the control delay. If this is the case with your equipment, don't let it stop you from trying AMTOR. There are some stations you won't be able to contact, but there are many other stations you *will* be able to contact—depending on propagation and equipment delays.

There is another programmable delay in most AMTOR controllers and MCPs: the *transmit delay* (TD). Transmitters don't reach full output instantly. Since this is the case, the transmit delay tells the MCP to hesitate before sending the data. I find that practically all HF transmitters work correctly if you set TD to 10 ms. This parameter can remain fixed since it will be valid for all AMTOR contacts. Of course, remember that your transmit delay—and the transmit delay used by the other station—will increase the total time delay in the ARQ link.

are too long, ISS equipment will still be switching when it should be receiving. As a consequence, it will never hear the control-character transmissions.

As I'm sure you've guessed, your AMTOR communications range is limited by distance, and the speed at which your equipment operates. You'll find, for example, that you can't seem to work stations that are located beyond a certain distance (measured in *thousands* of miles). You'll also discover that you have a *minimum* range as well. The point to remember is that there are a *lot* of AMTOR operators within range of your station—more than enough to keep you busy for a very long time! Besides, you may be able to make a few adjustments that will extend your maximum range, or reduce your minimum range. See the sidebar, "Flexible Timing."

Eavesdropping on AMTOR

Most MCPs include some sort of *listen* mode. The listen mode allows you to tune to an ARQ contact in progress and print the text *without* being part of the link. You need a bit of patience when using the listen mode since it doesn't include the error-correcting feature. You'll see plenty of errors and repeated characters, but at least you can follow the discussion.

Listen mode also works differently, depending on the MCP you're using. The listen mode in some controllers automatically senses and switches to ARQ or FEC. Other units monitor only ARQ transmissions in the listen mode and must be manually switched to the standby mode to monitor FEC signals.

Switch your transceiver to LSB and your MCP to the listen mode. Now hunt for a chirping ARQ signal. I suggest that you look for a strong signal between 14.070 and 14.080 MHz. (Twenty meters isn't the only band with AMTOR activity, but it's one of the most popular.) When you've tuned the signal correctly—remember to watch your indicator—you should see characters within 15 to 20 seconds. If not, try another signal. On some controllers, you may have to reset the listen mode to restart the synchronizing process. With practice, you should be able to copy an ARQ signal with ease. The listen mode may drop out of sync at times, especially when one station ends its transmission and switches from ISS to IRS. This is normal since the listen mode can only synchronize to one station at a time.

Now, find an FEC signal. At first, FEC may be hard to differentiate from RTTY. An FEC transmission includes special synchronizing characters, but they're often sent only once per line of text. It can take 10 seconds or more to receive these synchronizing characters—longer if you miss them or get a noise burst when they're sent. Some MCPs send extra FEC synchronizing signals and these signals will be easier to receive. Practice tuning FEC before you attempt your first AMTOR contact.

Let's Call CQ!

Switch your MCP to the FEC mode. (Some MCPs omit this step. They permit you to send FEC while in the ARQ mode.) An AMTOR CQ should include both your call sign *and* your SELCAL code (see the sidebar, "What's Your SELCAL?").

[switch your transmitter on]

[send a blank line]

CQ CQ CQ DE WB8IMY WB8IMY WB8IMY (WIMY WIMY WIMY)

CQ CQ CQ DE WB8IMY WB8IMY WB8IMY (WIMY WIMY WIMY)

CQ CQ CQ DE WB8IMY WB8IMY WB8IMY (WIMY WIMY WIMY) K

[switch back to receive]

The last step is *very* important! Your AMTOR controller must return to *standby* to be ready to receive an ARQ call. Most AMTOR controllers have two different ways to end an FEC transmission: one command returns the controller to the

What's Your SELCAL?

Before you can operate AMTOR ARQ, you have to choose your *selective call identifier*, or *SELCAL*. When AMTOR stations wish to communicate in ARQ, this is the code that must be used to establish the link. The SELCAL code uses *only* letters, and we choose letters that match at least part of our call signs. Some examples are:

Call Sign	CCIR-476 SELCAL	CCIR-625 SELCAL
KS9I	KKSI	KSIIXXX
WA9YLB	WYLB	WAIYLBX
W1AW	WWAW	WAAWXXX
WB8IMY	WIMY	WBHIMYX

(CCIR-476 is the name of the recommended technical specifications for the version of TOR that's most popular on the amateur bands today.)

Most amateurs use only the CCIR-476 SELCAL configuration. The letter combinations shown for CCIR-625 are strictly my own choice—you can use others.

Most MCPs feature a special command that will allow you to store your SELCAL in memory. (In the Kantronics KAM, for example, the command is "MYSEL." In the PK-232, it's "MYSELCAL.") This is an important step! Without a SELCAL, other AMTOR stations will not be able to call and link to you in the ARQ mode.

But Wait a Minute! What is CCIR-625?

CCIR-625 is a revised AMTOR/SITOR international standard. It was devised to address two problems: (1) the four-character CCIR-476 code was too limited to provide different SELCALs to all stations, and (2) under some circumstances, a CCIR-476 station could re-link with an incorrect station if the original link failed. CCIR-625 allows *seven* characters in its SELCAL, automatically identifies both stations at link-up, and also tightens the specifications for FEC synchronization.

Newer AMTOR controllers include both modes, but CCIR-476 is compatible with both new and old equipment. In a few years, however, CCIR-625 may become the dominant Amateur Radio AMTOR format.

standby mode and another returns to the FEC mode. Be sure to check your manual!

Also remember to send your FEC CQ with lots of short lines terminated with carriage returns. Why? It allows receiving MCPs to quickly synchronize with your signal. The faster they synchronize, the sooner they'll copy your message!

If someone wants to chat with you, they'll call you in ARQ using the SELCAL code you sent (WIMY in our example). If you don't get an answer, try again. Keep your calls short—don't get fancy or long-winded!

Let's assume that someone is answering your call. You hear the familiar chirp-chirp-chirp as they send your SELCAL over and over. In the meantime, your MCP is decoding the signals. Does the received SELCAL match yours? Yes! At this point you may hear a tone or chime followed by a message telling you that you're linked to the other station.

The AMTOR dance has begun, but who calls the tune? Well, a station that answers (in ARQ) the FEC call of another station is the *master*. It sets the timing parameters and the other station—the *slave*—must synchronize to its signals. In this case, the station that called you is the master and you're the slave. Throughout the conversation, the master and slave roles remain fixed.

Anatomy of an ARQ Conversation

Once an ARQ link begins, the exchange of ACKs and NAKs goes on *continuously*—even if no one is sending information. If you and the other operator decided to leave your keyboards and grab a snack, your stations would chirp mindlessly back and forth to each other. I wouldn't recommend this as a standard operating practice, though!

When a station calls you and establishes a link, you're the IRS. Just wait patiently for the other operator to send his greeting . . .

WB8IMY DE KU7G . . .
HELLO! MY NAME IS BOB AND I LIVE IN
WASHINGTON STATE IN THE SHADOW OF
MOUNT ST HELENS. YOUR RST IS 589.
BACK TO YOU . . . +?

Notice how Bob leads off with his call sign and yours. This is important since all you've exchanged so far are you SELCALs. He sends a brief greeting and ends it with +?. This strange-looking character is the *over* command. It allows the stations to trade places from ISS to IRS. Bob is turning the link over to you so you can reply.

Over to You

There is a very important point to remember about AMTOR ARQ: *each station must turn the link over to the other station at the end of every exchange.* (This has nothing to do with the master/slave timing relationship.) The normal procedure is to type +? to switch the link. With many MCPs, the +? is sent by tapping a single key. In effect, this control code says, "Let's turn the link around. You're the ISS now." When the link switches, you'll hear a distinct change in the chirping rhythm.

Recognizing that there are times when the IRS operator (receiving station) would like to immediately break in and make a comment, most MCPs include a *forced over* command. A forced over causes an immediate link reversal, even if the ISS operator is still typing or has text in his transmit buffer. The exact command used to cause a forced over varies between units. Use the forced over sparingly; it's rarely needed, but very handy at times.

When I made my first AMTOR ARQ contact, I didn't understand the idea of turning over the link. I sent my name, location and signal report, and then sat back and waited for a response. It wasn't long before I began to wonder why the other station wasn't answering. Our transceivers were chirping happily to each other, so everything *seemed* to be okay. I saw the letters "ISS" flashing in the corner of my screen, but I really didn't know what they meant! (Like many hams, I often dive right into a new mode without reading my manuals.) Finally, the other station realized that he had a greenhorn on his hands. He sent a forced over and began to patiently explain the meaning of ISS, IRS and why I had to turn the link over to him!

The Conversation Continues

So Bob has turned the link over to you. Now you're the ISS and he's the IRS. Why not send a short greeting and a signal report, too?

HELLO, BOB. YOU ARE SOUNDING FINE HERE IN CONNECTICUT. I LIVE IN A TOWN CALLED WALLINGFORD AND MY NAME IS STEVE. YOUR RST IS 599.

There is no need to repeat the information as you might do during a RTTY exchange. With the ARQ ACK/NAK system, the other station either receives your text or doesn't! With rare exception, the text is never garbled on the receiving end. If there is trouble on the frequency (due to fading or interference), the flow of incoming or outgoing text will slow down or stop altogether.

It sounds like Bob lives in an interesting place. Maybe you should ask him about it.

WERE YOU LIVING NEAR MOUNT ST HELENS WHEN IT ERUPTED BACK IN 1979? KU7G DE WB8IMY +?

Good! You remembered to turn the link over. Now Bob can respond.

YES, I WAS JUST MOVING INTO MY NEW HOME WHEN THE MOUNTAIN BLEW ITS TOP. THERE

WERE ASHES EVERYWHERE! I COULDNT DRIVE TO MY NEW JOB BECAUSE OF ALL THE ASH IN THE ENGINE. I STILL HAVE SOME OF IT IN JARS IN MY BASEMENT! +?

Now you have the start of a fascinating conversation! As with RTTY, you can type your comments while you're receiving his. When the link turns over, your system will begin sending the pretyped text automatically. All good things must come to an end, however...

WELL, STEVE, I HAVE TO GET UP VERY EARLY TOMORROW AND DRIVE ALL THE WAY TO BOISE, IDAHO. I THINK I SHOULD GET TO BED SOON OR I WILL NEVER HEAR THE ALARM CLOCK. THANKS FOR THE GREAT CONVERSATION. I REALLY ENJOYED IT. GO AHEAD AND MAKE YOUR FINAL COMMENTS AND THEN YOU CAN DOWN THE LINK. 73... WB8IMY DE KU7G SK +?

Bob says you can "down the link"? What does that mean? It simply means that he is asking you to send the *end* command that will terminate the ARQ link between your stations.

GOODNIGHT, BOB. HAVE A GOOD TRIP. HOPE TO LINK UP WITH YOU AGAIN ONE OF THESE DAYS. 73... KU7G DE WB8IMY SK

Most MCPs feature a single keystroke that sends the end command (some label it "disconnect"). If you're in doubt, you can do it manually by sending **ZZZZ**.

That's it! The link is broken and the conversation is over. If another station has been eavesdropping, he or she may call now. Only two stations at a time can be linked via AMTOR ARQ. If anyone else wishes to talk to you, they must wait until the conversation ends.

Answering an AMTOR CQ

What if you're prowling through the RTTY/AMTOR subbands and you see someone calling CQ? The first order of business is to load the *other station's* SELCAL code. Be careful not to change your own in the process! Obviously, there are two SELCAL codes involved: your SELCAL and the SELCAL of the other station. To help keep this straight, MCPs label your SELCAL as *MYCALL*, *MYA*, *MYSEL* or *LOCAL CALL* (LC).

Most MCPs will prompt you to enter the SELCAL of the other station:

SELCAL? ____

Some units (such as the Hal PCI-3000) include a call directory that lists several calls in a menu format. Still other controllers label the SELCAL to be sent as the *remote call* (RC), or sometimes *HISCALL*. Consult your manual to determine the correct label and the proper procedure to enter the other station's SELCAL.

Once you've entered the SELCAL, your station will begin chirping it over and over. If the link isn't established within a certain period of time, the transmissions stop. With luck, the other station will hear your signals and start the link. You're the calling station, so you're the master now. All the timing parameters for this conversation will be dictated by *your* station. Since you started the conversation, you're also the ISS at the moment. Send your opening remarks and start having fun!

APLink: Your Window on the Packet World

APLink is a bulletin board system (BBS) program designed by Vic Poor, W5SMM, to provide message store-and-forward capability for AMTOR users. APLink allows stored messages to be shared between two computer ports—one port for an HF AMTOR controller and another for a VHF

You don't need three computers to operate AMTOR, but KG5EG puts them all to good use. His neat station layout makes it easy for him to operate any mode he desires. Note the Alpha linear amplifier on the cabinet at the far right. It's rated for high output and fast transmit/receive switching.

packet TNC. Messages can be read or stored by an AMTOR station on HF, or by a packet station on VHF. As a result, APLink provides a connection (no pun intended!) between AMTOR operators and the VHF packet network.

The APLink Scanning BBS

Frequency-scanning APLink stations first appeared in Europe as the brainchild of G3PLX. They're now common in the United States and they're spreading worldwide.

A scanning APLink BBS uses the programmable memories and scanning options of newer HF transceivers. When the BBS is not in use, its receiver continuously scans the programmed frequencies. A BBS station often scans four or five bands and two or three different frequencies within each band. A complete scan can take 15 to 30 seconds.

The scan pauses for 2 to 5 seconds on each frequency. If the BBS station hears its own SELCAL being sent by a potential user, scanning ceases and an ARQ link is established. If the frequency is busy, or the BBS SELCAL is not heard, the BBS resumes scanning.

This frequency-scanning technique allows one BBS to serve stations at varying distances—regardless of propagation conditions. If you can't connect to an APLink BBS on 20 meters, for example, try again on another band. Eventually you'll find a band that offers a good path between you and the APLink BBS. It's a very clever way to make the best use of our available HF spectrum and propagation!

If you want to connect to a scanning APLink BBS, you must: (1) know the exact frequencies the BBS is scanning, and (2) be able to accurately set your transmitter to those frequencies. Scanning BBS stations have files that list the exact frequencies they monitor. When you first connect to an APLink BBS, download its frequency list so you'll know where to find it the next time. Table 4-1 shows a list of APLink stations in operation when this book was published.

Before attempting to use an APLink station, listen to several of the listed frequencies and see if you can hear the BBS communicating with other stations. This will tell you whether the BBS is busy (calling it will do no good if it is), and the quality of the BBS station's signal at your QTH. When you finally call the BBS, choose a frequency where its signal is strong and stable.

Most APLink operating commands will be familiar to packet users. As is the case with many packet BBS systems these days, APLink stations have disabled their command list response. This means that you may only see **GA +?** when you initially establish the link ("GA" means "Go Ahead").

Sending a private message via APLink is very similar to packet, but **AT** is used in place of @. For example, on an APLink BBS you'd enter: **SP WA0QKL AT W0AMN** rather than **SP WA0QKL @ W0AMN**. (Even this is changing with the advent of upper/lower-case AMTOR. These APLink stations *do* use @ in their message format—just like packet.)

Table 4-1

APLink Systems Worldwide

Note: All frequencies shown are MARK frequencies

Call sign	SELCAL	Location	Frequencies (kHz)
United States			
KB1PJ	KBPJ	Boston, MA	3622, 3625, 10126, 10128,
K1UOL	KUOL	Bethel, CT	14071.5
N2JAW	NJAW	Trenton, NJ	3622, 7071, 7072.5, 7075.5, 10128, 10140.5, 14068, 14071.5, 14073.5, 21072.5
W2NRE	WNRE	Scarsdale, NY	3820, 7068, 7070, 7072, 7074, 10126, 10128, 10140, 14070, 14072
N3EXW	NEXW	Rockville, MD	14070.5
W3GL	WWGL	New Castle, DE	7071, 7072.5, 7075, 7075.5, 14068, 14069, 14071.5, 14073.5
KK4CQ	KKCQ	Pensacola, FL	3622, 7070, 7070.5, 7071, 7076, 10126, 10128, 14070, 14071.5, 14072.5, 21074, 21076, 28070, 28128
K4CJX	KCJX	Nashville, TN	3622, 7068, 7070.5, 7072.5, 7075.5, 10126, 10128, 10139.5, 14068, 14069.5, 14070, 14072, 14076, 21074, 21076, 28128
W4NPX	WNPX	Charlottesville, VA	7068, 7070, 7070.5, 7072, 7074, 10126, 10128, 10140, 14070, 14072, 14074
K4YZU	KYZU	Louisville, KY	7069, 7071, 14069.5, 14071.5, 10140.5, 10141.5, 14079.5, 21072.5
W2TKU/4	WTKU	Sarasota, FL	3622, 7070, 7072, 7076, 10126, 10128, 10140, 14066, 14068, 14070, 14076, 14078, 21072, 21074, 21080
K5CVD/4	KCVD	Windsor, SC	7070, 7072, 7074, 10126, 10128, 70140, 18102.5, 18105.5, 24915, 24925

continued on next page

Call sign	SELCAL	Location	Frequencies (kHz)
WA9FCH/4	WFCH	Reston, VA	7070.5, 7071, 7072.5, 7075.5, 10128, 10139.5, 10140.5, 10140, 14068, 14070.5, 14071.5, 14072.5
KP4GE	KPGE	Caguas, PR	14066, 14067, 14068, 14069, 14070, 14071, 14072, 14073, 14074, 14075, 14076, 14077, 14078, 14079
KE5HE	KEHE	Hearne, TX	3622, 7068, 7069, 7071, 10126, 10128, 10139.5, 14070.5, 14071.5, 14072.5, 14079.5 21072.5 24925 28125
W5KSI	WKSI	New Orleans, LA	14070, 14070.5, 14073.5, 14074, 14075, 14077, 14079, 14080.5 Day/Night; 21081, 21079, 21075, 21074, 28074, 28075 Day only.
NZ2T/5	NNZT	Dallas, TX	7071 (2AM-10AM Central Time) 7069, 7071, 7073, 10129, 10131, 14069, 14071, 14073, 14075, 21073, 21077 24915, 28077, 28129
WB5UJO	WUJO	Marlin, TX	7075
WA8DRZ/6	WDRZ	San Francisco, CA	10128, 10129, 14068.5, 14069.5, 14070.5, 14071.5, 14072.5, 14073.5, 14074.5, 14075.5
NH6VT	NHVT	Waialua, HI	14069, 14070.5, 14071.5, 14072.5, 14075, 21076, 21079, 28074
AA5CQ/7	AACQ	Las Vegas, NV	10140.5, 14070.5
AA7HS	AAHS	Yakima, WA	7069 or 7071 Evenings, 14.072.5 Days
K7BUC	KBUC	Phoenix, AZ	7071, 10140, 14071.5, 14073.5, 14074, 21073.5 Days; 3627, 7071, 10140, 14071.5, 14073.5, 14074, Evenings
K7SLI	KSLI	Marysville, WA	7073 Days, 3629 Evenings
KC7J	KKCJ	Tacoma, WA	14069

Call sign	SELCAL	Location	Frequencies (kHz)
KD7UM	KDUM	Salt Lake City, UT	3623, 3627, 7073, 7075, 10127, 10141, 14069, 14073, 14077, 21071, 21075, 28075, 28127
NØIA/7	NNIA	Las Vegas, NV	3625, 3627, 7069, 7071, 7072.5, 10128, 10139.5, 10140.5, 14070.5 14072.5 21072.5, 21074, 28070, 28128
N6EQZ/7	NEQZ	Seattle, WA	3605.2, 3629, 7071, 7073, 10126, 14068, 14069, 14071, 14073, 14075
N7CR	NNCR	Spokane, WA	3622, 7069, 7075.5, 10126, 10128, 14070.5, 14072.5, 18105.5, 21072.5, 21079, 24915, 28073
NA7P	NNAP	Seattle, WA	14069
WØLVJ/7	WLVJ	Spanaway, WA	3605.37
W2USA/7	WUSA	Fort Lewis, WA	28147.9
W5VBO/7	WVBO	Phoenix, AZ	3622, 7069, 10126, 10128, 14070.5, 14071.5, 14072.5, 18102.5, 18105.5, 21072.5, 24915, 28074
W7DCR	WDCR	La Pine, OR	3622, 7069, 7075.5, 10126, 10127, 10128, 14069, 14070.5, 14072.5, 18105.5, 21072.5, 21076, 21079, 24915, 24925
WI7D	WWID	Las Vegas, NV	3621, 3623, 7069, 7075, 10129, 10137, 14069, 14073, 14077, 18099, 21073, 21075, 24925, 28073, 28077, 28125
AL7LS	ALLS	Delta Junction, AK	14072.5
W7IJ/8	WWIJ	Cleveland, OH	3622, 7071, 10126, 10128, 14066, 14069, 14071.5, 18105.5, 21070
W9MR	WWMR	Keensburg, IL	3622, 7070, 7074, 7076, 10128, 10140, 14068, 14070, 18104, 21072, 21074, 24925, 28128
WA1URA/9	WURA	Ft Wayne, IN	3622, 7071, 7075.5, 7076.9, 10128, 10139.5, 10140.5 14068, 14069, 14070.5, 14071.5, 14073.5, 14075, 21076, 21079, 28076.5

continued on next page

Call sign	SELCAL	Location	Frequencies (kHz)
WA9WCN	WWCN	Indianapolis, IN	3620, 3622, 7072, 7074, 10126, 10128, 10140, 14066, 14068, 14070, 14076, 21072
KAØJRQ	KJRQ	Omaha, NE	3622, 7075.5, 7071, 10126, 10130, 14071.5, 14074, 14072.5, 18105.5, 21074, 21071.5, 24915, 28.074

International

Call sign	SELCAL	Location	Frequencies (kHz)
9K2DZ	KKDZ	Kuwait City, Kuwait	7071, 10128, 14066, 14070, 14074, 14076.5, 14079, 21076, 21076.5, 21079, 18105.5, 24925
9K2EC	KKEC	Kuwait City, Kuwait	14071, 14072, 14079, 18102, 18105.5, 21071, 21081, 24925, 28079
9X5L	XXLJ	Kigali, Rwanda	14071, 14073, 14074, 14075, 14078, (1500-0500 UTC); 21071, 21073, 21074, 21075, 21078, (0600-1500 UTC)
BV2BV	BVBV	Taipei, Taiwan	14069
BV5AF	BVAF	Taipei, Taiwan	14072
BV5AG	BVAG	Taipei, Taiwan	21072
CE3GDN	CGDN	Santiago, Chile	21074
DKØMHZ	DMHZ	Hamburg, Germany	14081, 14081.5, 14082, 14082.5, 14083, 14083.5
DKØMUN	DMUN	Munich, Germany	14076, 14078, 14079, 14081, 21079, 21080, 21081
DU1AUJ	DAUJ	Quezon City, Philippines	14070 (1300-2300 UTC) 21070 (2300-1300 UTC)
DU9BC	DUBC	Davo City, Philippines	7012.8 (2300-1000 UTC) 14072 (1000-2300 UTC)
DU9WX	DUWX	Iligan City, Philippines	7012.8
FK8BK	FKBK	Noumea, New Caledonia	14066 (0700-1300 UTC)
GB7EMX	GEMX	Aberdeen, Scotland	3587.5, 3588.5, 3589, 7038, 7039, 7040, 10145, 10146, 14075, 14076, 14077, 14078, 21080, 21081
GB7PLX	GPLX	Portsmouth, UK	3588.5, 3589, 7038, 7039, 7040, 10145, 10146, 14076, 14077, 14078, 21080, 21081, 28075

Call sign	SELCAL	Location	Frequencies (kHz)
GB7SCA	GSCA	Plymouth, UK	3587.5, 3588.5, 3589, 7038, 7039, 7040, 10145, 10146 14075, 14076, 14077, 14078, 21080, 21081, 28075
GB7SIG	GSIG	Dorset, UK	3587, 3587.5, 3589, 7035, 7036, 7037, 10140, 10141, 10146, 14076, 14077, 14078, 21081, 28075
HB9AK	HBAK	Zurich, Switzerland	3581, 3583, 3588, 3589, 7038, 7040, 10141, 10146, 14071.5, 14072, 14075, 14078, 21080, 21085, 28075, 28080
HC5K	HHCK	Cuenca, Ecuador	210/4 or 28047
HL9TG	HLTG	Camp Humphreys, Korea	10140.5, 10146, 14069, 14069.5, 14070.5, 14071.5, 14072, 14072.5, 14073.5, 14074.5, 14075.5, 14077, 14079.5, 28074, 28125, 28128, 28147.9
IKØNNT	INNT	Rome, Italy	14078
JA1JTA	JJTA	Sagamihara, Japan	14070 (Sat/Sun only)
JA5TX	JATX	Kochi, Japan	14071, 14072, 14074, 14076, 14078 (PacTOR and AMTOR)
LZ2BE	LZBE	Razgrad, Bulgaria	7038 (2000-0600 UTC) 14074 (0600-2000 UTC)
OE4XBA/4	OXBA	Eisenstadt, Austria	21075.5 (0600-1800 UTC) 14075 (1800-2100 UTC)
OH2BAW	OBAW	Helsinki, Finland	3581.5, 3587, 7038, 10146, 14071, 14076, 21077.5, 28077.5
ON6RU	ONRO	Louveigne, Belgium	14074, 14076 (1600-2400 UTC) 21074, 21076 (0700-1600 UTC)
PAØRVR	PRVR	Papendrecht, Netherlands	14068, 14069, 14070, 14071, 14072, 14074, 14075, 14078
SL5BO	SLBO	Stockholm, Sweden	14077
SM4CMG	SCMG	Fellingsbro, Sweden	3581.5, 7038, 7039, 7040, 10145, 10146, 14074, 14075, 14076, 14077, 18105.5, 21074, 21075, 21076 (Mon/Fri only)

continued on next page

Call sign	SELCAL	Location	Frequencies (kHz)
SM6FMB	SFMB	Gotenburg, Sweden	7037, 7038, 7039, 10109, 10128, 10145, 10146, 14069, 14070, 14075, 14076, 14078, 18102.5, 18105.5, 21072, 21074, 21075, 21076, 28074
SU1ER	SUER	Cairo, Egypt	14066 (1600-2000 UTC) 21070 Fri/Sat (0800-1000 UTC)
TG9VT	TGVT	Guatemala City, Guatemala	7068.5, 10128, 14066, 14068, 14069, 14074, 18105.5, 21070, 21072, 21074, 24915, 28074
TU2BB	TUBB	Abidjan, Ivory Coast	14076 Night, 21076 Day (Mon/Fri only)
TY1PS	TYPS	Cotonou, Benin	14072, 14078, 21072, 21078, 28072, 28078
U5WF	UUWF	Lvov, Ukraine	14075
V51NH	VVNH	Windhoek, Namibia	14070
VE3PAO	VPAO	Toronto, ON, Canada	7070 (0000-0400 UTC) 14070 (0400-2400 UTC)
VE6PD	VEPD	Lethbridge, AB, Canada	10126, 10128, 10139.5, 10140, 14069, 14070.5, 14076, 21072.5, 21076, 28075
VE7CTJ	VCTJ	Squamish, BC, Canada	7072 14072.5
VE7DYT	VDYT	Port Alice, BC, Canada	7073 (0700-0300 UTC) 3629 (0300-0700 UTC)
VO1BBS	VBBS	Seal Cove, NF, Canada	14068.5
VK2AGE	VAGE	Lismore, Australia	7045, 10109, 14075, 14077, 21076

I use APLink to send mail to a packet-active friend in Dayton, Ohio—about 1,200 miles from where I live in Connecticut. Sure, I could send the message via the regular packet network, but it might take several days to reach him. APLink offers a time-saving shortcut! I link with the WA1URA/9 APLink in Fort Wayne, Indiana—not far from Dayton in terms of packet networking. After I enter my message, the APLink uses VHF to transfer it to my friend's

Call sign	SELCAL	Location	Frequencies (kHz)
VK2CBF	VCBF	Sydney, Australia	14069
VK2EHQ	VEHQ	Sydney, Australia	14070.5
VK2OG	VKOG	Sydney, Australia	14069
VK3WZ	VKWZ	Melbourne, Australia	14075
VU2DPG	VDPG	New Delhi, India	14079 (1600-0300 UTC) 21079 (1300-1600 UTC) 28079 (0300-1300 UTC)
AA6VY/XE2	AAVY	Punta Banda, Mexico	21080 (1600-1900 UTC) 14078 (0400-0700 UTC)
ZF1GC	ZFGC	Bodden, Grand Cayman	14070.5, 14071.5, 14072.5, 14073.5, 14074.5, 14075.5, 14076, 21080
ZL1ACO	ZACO	Auckland, New Zealand	10128, 14070.5, 14072.5, 14073.5, 14075, 14075.5, 21076, 21079
ZL4AK	ZLAK	Oamaru, New Zealand	14069, 14070.5, 14072, 14074, 14075, 14077, 21074, 21076, 21079, 28074
ZS5S	ZZSS	Howick, South Africa	7037, 14069, 14073, 21069, 21073, 28073
ZSQYKM	ZSKM	Pretoria, South Africa	14075 (0500-0700 UTC) 21075.5 (1500-1700 UTC) 14075 (1700-1900 UTC) (Mon/Fri only)
ZS6ZQ	ZSZQ	Johannesburg, South Africa	3586 (1600-0500 UTC) 7038 (0500-1600 UTC)

(*Information supplied by WA8DRZ*)

local packet bulletin board, usually within one hour! He can send his reply on VHF back to the WA1URA/9 APLink. The next time I establish a link to the system, it will tell me that I have mail waiting.

Let's Give it a Try!

What you're about to see is text from an actual link with the WA1URA/9 APLink system. I select 14.071.4 MHz (suppressed carrier frequency) and begin calling by sending the SELCAL **WURA** in the ARQ mode. Within seconds, we're linked!

My transmissions are shown in italics. Note that I don't need to use the **+?** command to turn the link over. APLink expects each command to be followed by a carriage return and a line feed. When it senses this, it sends a forced over to switch the link automatically.

DE WA1URA/9 APLINK 6.MT STANDBY

NEED HELP? TYPE 'HELP' (CR)

... PLEASE LOGIN+?

LOGIN WB8IMY

WB8IMY DE WA1URA/9 QRU GA+?

("QRU" means there's no mail waiting for me. If I *did* have mail waiting, I'd see "QTC" followed by the message number. Before I proceed, I ask for a list of recent users.)

LR

USERS IN THE LAST 24 HOURS:

9K2DZ	**GB7SCA**	**KA0JRQ**	**KB1PJ**	**KB8NH**
KD4B	**KE5HE**	**KK4CQ**	**N0IA**	**N2JAW**
NR0S	**UA9QRA**	**TG9VT**	**VE6PD**	**W7DCR**
W7IJ	**WA4DDX**	**WB3EPC**	**WB8IMY**	**WG1I**
ZS5S				

GA+?

(I think I'll perform a test by sending mail to myself! I'll be curious to see how long it takes for the message to arrive at my local packet BBS in Connecticut.)

SP WB8IMY AT W1NRG.CT.USA.NA

CONFIRM ... SP WB8IMY AT W1NRG.CT.USA.NA (YES/NO) +?

> ### WRU Answer-Back
> Most AMTOR controllers and all commercial SITOR controllers include an automatic station identification feature called "Who Are You" (WRU). This feature dates back to mechanical Teletype machines. The concept involves sending a special character code that triggers an automatic response from another station—usually its call sign.
>
> AMTOR WRU works like this:
>
> 1. My WRU feature is active and I have text in my *ANSWERBACK* storage (my call sign).
> 2. Your station is the ISS and you send the WRU code.
> 3. My AMTOR controller responds with a *forced over* and becomes the ISS.
> 4. My station sends its ANSWERBACK text followed by *over* (+?).
>
> Note that it should be the station called (mine in this example) that forces the first *over*. If your MCP doesn't handle the exchange in this manner, using your WRU feature can create some *very* confusing situations! For this reason, you may not care to use the WRU function unless you are also using APLink.

YES

MSG NR 61497 GA SUBJ/MSG+?

(This is the tricky part for new users. The APLink wants me to enter the subject of the message *on a line by itself*. After that, I'm free to go ahead and enter the rest of the message.)

TEST MESSAGE
This is an APLINK test message being sent back to my home packet PBBS.
NNNN

(Notice how I ended the message with NNNN, also on a line by itself.)

Time to Start Chirping with AMTOR

Table 4-2

Common APLink Commands

- ❑ Commands to APLink on its AMTOR port should always be on a new line and end with CR.
- ❑ The user should avoid transmitting the "+?" sequence. APLink knows when it is its turn to send. APLink will change the link direction, as needed.
- ❑ Arguments shown in square brackets [like this] are optional. Arguments shown in point brackets <like this> are not optional and must be included.

A—Abort the current output; return to GA+? prompt
CANCEL <num>—Cancels message <num> if originated by you
F—Abort the current msg
LOGIN <call>—Logs you into the system
LOGOFF—Same as LOGOUT
LOGON <call>—Same as LOGIN
LOGOUT—Logs you off
H—Send the help file
L—List all non-bulletin, non-private messages
L [number]—List all non-bulletin, non-private messages equal to or greater than (number)
LTO or **LM**—List all messages to you
LTO [call]—List all messages to <call>
LFM—List all messages from you
LFM [call]—List all messages from <call>
LB—List new general interest bulletins
LB [number]—List general interest bulletins from [number] and higher
LT—List all NTS messages
NTS—List all unforwarded NTS messages (may be restricted)
LU—List all registered users
LR—List all stations that have logged on during last 24 hours
RN or **RM**—Read all new messages addressed to you
R [number]—Read message [number]
RH [number]—Read message [number] including routing headers
RI—Read the Intercept File (for forwarding to home BBS)
RF— Read the Auto-Fwd File (to see how AMTOR routing is done)
SP <call>—Send a private message to <call>, end with NNNN
SP <call1> AT <call2>—Send a private message to <call1> to be forwarded by packet to BBS with call sign of <call2>
ST <zip>AT NTS<st>—Enter (Send) an NTS message
SB [name]—Enter (Send) a bulletin to "name"; End with NNNN
SB [name1] [AT name2]—Enter a bulletin to be forwarded; End with NNNN
T—Talk to the SYSOP
V—Read version number
/// Anywhere on a command line cancels the command

WB8IMY DE WA1URA/9 NR 61497 FILED GA+?

(My message has been filed in the system as message number 61497. While I'm linked, I'll check for any NTS traffic to be delivered.)

LT

MSG	TS	SIZE	TO	AT	FROM	FILED(Z)	SUBJECT
61490	TY	474	NTSMN	NTSMN	W7DK	0913/1340	GILBERT MN (218)749
61489	TY	468	NTSIL	NTSIL	W7DK	0913/1338	MIDLOTHIAN IL (708)687
61474	TN	502	38346	NTSMS	N1JBR	0913/1005	GREETING
61472	TY	517	93550	NTSCA	WA1GDJ	0913/0957	PALMDALE 805-947
61444	TY	331	96703	NTSHI	N2JAW	0912/2352	WELFARE - KAUAI HI (808)826
61418	TN	599	63108	NTSMO	N4UAV	0912/1724	ST LOUIS MO 314-534
61410	TY	582	55313	NTSMN	N1MQR	0912/1027	QTC BUFFALO (612-477)

WB8IMY DE WA1URA/9 GA+?

(It doesn't look like there's anything waiting for delivery to my area of the country. I'll leave the APLink for now and check back in a few days.)

LOGOFF

WB8IMY DE WA1URA/9 SK

In case you're wondering, my test message arrived at the W1NRG packet bulletin board 6 hours later!

In addition to sending messages, you can read informative bulletins and access other information. Once you feel confident with AMTOR, try working an APLink system! See Table 4-2 for a list of common APLink commands.

CHAPTER 5
Exploring PacTOR and CLOVER

The HF bands are rugged territory for modern digital communications. High levels of natural and manmade noise are common. Propagation can shift from miraculous to abominable in a heartbeat. Interference from other signals is a constant headache.

So why would any reasonable human being try to pass digital information on the HF bands at all? Why not use VHF, UHF or microwaves instead? There's much more room, noise levels are low and propagation is relatively stable.

While the bulk of high-speed digital communications has indeed moved to frequencies above 30 MHz, the challenge of HF continues to attract the innovators among us. All over the world these amateurs constantly ask themselves, "Is it possible to transmit and receive digital information at acceptable speeds, without errors, on HF frequencies?"

Why not use RTTY? Nice try, but no cigar! Its limited number of characters and inability to detect errors knock it out of contention. AMTOR offers an improvement, but it suffers from slow *throughput* (the rate at which data travels from one station to another) and the same truncated character set. Also, AMTOR's error-detection capability falls just short of 100%. For reliable data communications, anything less than 100% is unacceptable.

That leaves HF packet in the winner's circle, right? Not really. While packet supports the full ASCII character set with 100% error-detection capability, throughput is another story. Unless reception is excellent at both ends of the path, the throughput of HF packet is far less than AMTOR. Throw in some noise, fading signals or interference and its performance level drops like a brick.

If RTTY, AMTOR and packet don't fit the bill, what will? Ask most digitally active hams and they'll say they want an HF mode that features. . .

❑ The complete ASCII character set and the ability to support binary data transfers

❑ The ability to send and receive data without errors

❑ Acceptable performance under virtually all signal conditions

❑ Minimal signal bandwidth

❑ Operation with existing transceivers

❑ Reasonable cost

That's a tough set of requirements—so tough that some have called it impossible! Of course, the first step to solving a sticky technical problem is to say it's "impossible." That pronouncement alone will send hundreds of amateurs into a frenzy, working long hours for nothing more than the satisfaction of proving otherwise!

Several years of effort have resulted in two new HF digital modes. Both appeared in the early '90s, and both hold the potential for solving the HF problem.

The AMTOR/Packet Fusion: PacTOR

PacTOR is the brainchild of two German amateurs: Hans-Peter Helfert, DL6MAA, and Ulrich Strate, DF4KV. They introduced this mode to the Amateur Radio community in the late '80s. European hams were among the first to embrace PacTOR, but by 1993 the mode had spread to the United States and elsewhere.

Like packet, PacTOR transmissions utilize the full ASCII character set. This means that you can send upper- and lower-case letters as well as the full range of standard English punctuation. In addition, PacTOR can also support the transmission and reception of binary data. (Computer software and various types of files must be sent as binary data.)

PacTOR has a lot in common with AMTOR, too. Each block of information is sent at fixed time intervals and acknowledged with brief control signals. This gives a PacTOR signal its own distinctive *chirping* rhythm. Like AMTOR, there are PacTOR *master* and *slave* stations. Each sending station must turn over the link so that the other station can respond (sound familiar?).

When it comes to transmit/receive switching speeds, PacTOR is somewhat more liberal than AMTOR. An AMTOR station sends a block of characters and waits up to 240 ms for a reply (an ACK or a NAK). PacTOR, on the other hand, waits 320 ms. This extra 80 ms is critical for communication over long distances. Any SSB transceiver capable of switching from transmit to receive within 130 ms can be used for PacTOR. This encompasses virtually all rigs made since 1970—if not earlier.

PacTOR has the capability to communicate at varying speeds according to band conditions. Under good conditions, PacTOR will accelerate to 200 baud. It can go even faster (up to 400 baud) by compressing text data through the use of the *Huffman code*. If conditions deteriorate, however, PacTOR will automatically slow back down to 100 baud. You can switch speeds manually or, as most PacTOR operators prefer, allow the system to switch by itself.

Working a PacTOR Mailbox

While most PacTOR conversations take place between two human operators, there are a number of mailboxes on the air, too. If you remember our discussion of AMTOR APLink systems in Chapter 4, much of what you're about to see will look familiar.

You access a PacTOR mailbox as you would any PacTOR station: by sending its *complete* call sign (there are no SELCALs in PacTOR). If I want to connect to the W8KCQ mailbox, I use the PacTOR *call* command.

CALL W8KCQ

My controller keys my transmitter and attempts to establish communication with W8KCQ. If it's successful, I'll see:

*** NOW CALLING W8KCQ

*** CONNECTED TO W8KCQ

There is no need to log in to the BBS. It knows who I am already.

WB8IMY, QRA W8KCQ: (mailbox)

No messages on file for you
Send a command: (help) (cr)
Next?
=>

I think I'll ask for a list of mailbox commands by simply sending **COMMANDS**. Note that the mailbox switches the link automatically—just like an AMTOR APLink system.

WB8IMY, QRA W8KCQ:
COMMANDS: (02/11/92)

ABORT, BYE, EXIT, QRT OR QUIT - (Aborts current
   session)
ALARM - (Calls the system operator)

AUTOMBX - (Returns all of your mail and deletes it, ID and link down)
COMMANDS - (Returns this command list)
CONVERT (data) - (Converts upper case to lower case and vice versa)
DATABASE (call sign)- (Returns database for call sign)
DELETE (file name) Deletes indicated file (note 1)
DIR (call sign) - (Returns message directory for call sign, see SDIR)
DOWNLOAD - (Returns any stored messages for you)
DUMP - (Directs your following text to the system operator)
FILES - (Your message file status)
HAM NEWS-(Returns Ham news items or list of Ham news items)
HELP - (Help for using system)
INFO - (Returns information directory)
INPUT (call sign) - (Stores a message file to call sign)
LOG - (Returns a list of the last users)
OUTPUT - (Returns any stored messages for you)
PROTO - (Toggles System/User control protocol)
REACCESS - (call sign) - (Allows you to sign on again in case you got signed on incorrectly)
READ (file name) - (Returns stored file)
SDIR - (Short form message directory of all messages in the mailbox)
STATUS - (Returns system status)
STORE (call sign) - (Stores a message file)
STORE TEMP - (Stores a INFORMATION FILE for the SYSOP to place into HAM NEWS, etc.)
TRAFFIC - (Returns a condensed list of message files waiting pickup)
XFEC (file name) - (While in the ARQ mode, will return the listed file via the FEC mode. You must be listed in the mbxs' data base for the system to obtain your AMTOR call sign).

A XFEC (file name)/D Will cause the link to release after the file is transmitted.

(eof)
Next?
=>

Whew! What a list! I'll save that to disk for future use. Let's use the INFO command to see what files are available.

Note: The latest is listed first.

File Name	- Bytes -	Description
BMKPSW	- 2398 -	BMK-Multy PacTOR Software by G4RIA 10/13/92
PACINTRO	- 17.7k -	Introduction to PacTOR by W9UWE 10/01/92
USE15M	- 0583 -	Use/try 15m by KN6O
VE7CIZ.INF	- 0742 -	Information file for VE7CIZ 9/25/92
PROCOMM.HLP	- 8954 -	PACCOMM-PC+ Corrections/information by N4TEB 8/20/92
DJ2HZ.TP	- 6411 -	Description of Terminal Program by DJ2HZ
PACTOR.LST	- 0823 -	List of PacTOR CALLS and FREQUENCIES
PTHELP	- 0689 -	PacTOR Help (New Commands)
AMHELP	- 1796 -	AMTOR Help
PTCINFO	- 1237 -	Short writeup on PacTOR
AIRGRD	- 2460 -	SSB HF LIST OF AIRCRAFT TO GROUND FREQUENCIES
QRG	- 0302 -	SCAN/FREQUENCY LIST
MBXOP	- 1395 -	HOW TO USE THE MAILBOX SYSTEM

Do a command: READ (file name) to obtain any of the above.

TURN YOUR PRINTER ON?

(To CANCEL reading a file, do an over and send six X's (eof)

Next?
=>

Looks like another block of information to print and save! Before I leave this mailbox, I'll use the DUMP command to leave a message for the SysOp.

THIS IS WB8IMY, WALLINGFORD, CONNECTICUT
NAME: STEVE
EQUIPMENT: KENWOOD TS-820S
           PACCOMM PACTOR CONTROLLER
           IBM-XT PC
           DIPOLE ANTENNA UP ABOUT 30 FEET

THIS CONNECTION TO YOUR MAILBOX IS MY *FIRST* PACTOR CONTACT.
THANKS AND 73 . . . . STEVE
NNNN

0364 BYTES STORED IN FILE 64W8KCQ
Next?
=>

Now it's time to say "goodbye" . . .

BYE

WB8IMY, QRA W8KCQ CLEAR.
*** DISCONNECTED / QRT RECEIVED AT 11:31:08 - SUN, 01/11/92

Memory ARQ

PacTOR handles the challenge of sending error-free data in an interesting way. As with AMTOR, each character block is sent and acknowledged if it's received intact. If signal fading or interference destroy some of the data, a NAK is sent and the block is repeated—just like AMTOR. The big difference, however, involves *memory*.

When a PacTOR controller receives a mangled character block, it analyzes the parts and temporarily memorizes whatever information appears to be error-free. If the block is shot full of holes on the next transmission as well, the controller quickly compares the new data fragments with what it has memorized. It fills the gaps as much as possible and then, if necessary, asks for another repeat. Eventually, the controller gathers enough fragments to construct the entire block! This is a tremendous improvement over AMTOR and packet where the character block is often repeated many times before it finally makes it through unscathed. PacTOR's *memory ARQ* feature dramatically reduces the need to make repeat transmissions of damaged data. This translates into much higher throughput.

What Do I Need to Run PacTOR?

Assembling a PacTOR station is very simple. All you need are the following:

❏ An SSB transceiver

❏ A data terminal or a computer running terminal software

❏ A PacTOR controller, or an MCP with PacTOR capability

When assembling your PacTOR station, use the same guidelines discussed in Chapter 2.

PacTOR controllers are priced at about the same level as standard MCPs. In fact, many PacTOR controllers offer

The PacComm PacTOR controller. This model will operate RTTY and AMTOR *in addition* to PacTOR. It includes a PacTOR mailbox so that your friends can pick up and deliver messages when you're not in the shack.

RTTY and AMTOR *in addition* to PacTOR. Some MCP manufacturers are also adding PacTOR to their inventory of modes. See the RTTY/AMTOR Resource Guide for addresses of PacTOR equipment suppliers.

The future for PacTOR looks very promising. It meets all the requirements to become a popular HF digital mode. There is another digital contender in the ring, though. Despite its funny-sounding name, it has the ability to outperform even PacTOR!

I'm Looking Over a 4-Tone CLOVER . . .

Try designing a digital modulation scheme that uses phase-shift keying, amplitude-shift keying and a sequence of *four* tone pulses. What would you call it? If you were looking for a metaphor, the four tone pulses could be compared to . . . the leaves of the fabled four-leaf clover! Ray Petit, W7GHM, had a similar idea in mind when he developed the new HF digital mode known as *CLOVER*.

As this book went to press, the first CLOVER controllers

PacTOR Software

You can use any terminal program to "talk" to your PacTOR controller, but some are easier to use than others. Be sure to check all your software settings before getting on the air. For example, is the *word-wrap* or *line-wrap* on? This will add an automatic carriage return and line feed when the received text reaches the right-hand edge of your screen. Are you set for *full duplex* or *half duplex*? Half duplex works best with some PacTOR controllers. With practice you'll discover how to configure your terminal software for best results. When you do, save the settings to disk if possible. Some terminal programs maintain *configuration files* and will automatically use the settings you've saved.

Specialized PacTOR terminal software is also available. Werner Zielke, DJ2HZ, has designed a program to make it easier to use PacTOR controllers with IBM PCs and compatibles. This terminal software package is custom-designed for your station. The most recent versions even provide software control of your transceiver, if your rig has that feature.

If you're interested in Werner's software, you can contact him at the following address:

Werner S. Zielke, DJ2HZ
Im Winkel 13
D-2055 Dassendorf/Hamburg
Germany

For all information requests, include a self-addressed air-mail envelope and $2 for return postage. If you'd like to purchase his customized PacTOR terminal software, you must provide the following:

- A self-addressed *diskette* mailer
- A blank, formatted diskette (3½" or 5¼")
- $40 US. See your local bank or post office for information about sending money overseas.
- Your call sign
- Your AMTOR SELCAL
- The make and model of your transceiver

were being manufactured and sold by HAL Communications. The performance of the CLOVER system on the air has been nothing less than spectacular. In terms of throughput, CLOVER boasts the potential to outperform AMTOR by a factor of 10:1, if not more. Not only that, it can deliver this performance under truly terrible signal conditions.

How Does CLOVER Work?

CLOVER relies on digital signal processing to perform most of its magic. For example, CLOVER has the ability to correct most data errors *without* requesting repeat transmissions. A total of 31 damaged bytes can be repaired automatically out of every 188 bytes received. A request for a repeat transmission will be sent *only* if the number of flawed bytes exceeds this limit.

Like PacTOR, CLOVER is also adaptive, changing transmission speeds as conditions warrant. CLOVER takes this adaptive technique a step further, however. The CLOVER demodulator actually measures the signal-to-noise ratio, frequency offset and phase dispersion of *every* block of received data. As signal conditions deteriorate, the data rate is reduced. As conditions improve, the rate increases. By using this adaptive technique, CLOVER squeezes maximum efficiency out of every HF signal path.

CLOVER Handshaking

As you may recall, PacTOR and AMTOR both use an *over* command to switch the link so that one station can send while the other receives. CLOVER links must be switched as well, but the switching takes place *without* using *over* commands.

When two CLOVER stations make contact, they can send limited amounts of data to each other (up to 30 characters in each block) in what is known as the *chat mode*. If the amount of data waiting for transmission at one station exceeds 30 characters, CLOVER automatically switches to the *block data mode*. The transmitted blocks immediately become larger and are sent much faster. The other station, however, remains in the chat mode. Because of precise frame timing, all of this takes place without the need for either operator to change settings. And what if both stations have large amounts of data to send at the same time? Then they *both* switch to the

---

CLOVER Bandwidth

As we've already discussed, space is at a premium in the HF digital subbands. That's why it's important for any HF digital mode to be as *narrow* as possible.

With a 2-kHz bandwidth, a packet signal is nearly as wide as a voice transmission. You can't squeeze too many packet signals onto the band before serious interference begins (just listen to the HF packet activity on 20 meters!).

AMTOR improves the situation with a 1-kHz bandwidth, but that's still awfully wide for the available space. PacTOR transmissions have about the same bandwidth. If we expect to use the HF bands for efficient digital communications, we need to provide room for many more signals.

And then along comes CLOVER! As remarkable as it may seem, CLOVER manages to conduct extremely efficient communications while using only 500 Hz of spectrum! Two CLOVER signals could occupy the space of one AMTOR signal. *Four* CLOVER signals could fit in the same amount of spectrum required for one packet signal.

This narrow bandwidth is yet another prominent feature of CLOVER. Of course, a narrow-bandwidth signal requires more careful tuning. That's why the HAL *PC-CLOVER* software includes an on-screen tuning indicator. You can also understand why your transceiver must be very stable to operate CLOVER. With a 500-Hz wide signal, all it takes is a little bit of drift and you're way out of the ballpark!

data block mode. This high degree of efficiency is transparent to you, the operator. All you have to do is type your comments or select the file you want to send—CLOVER takes care of everything else!

CLOVER features an FEC mode similar to AMTOR. You use the CLOVER FEC to send transmissions that can be received by several stations at once. (In the CLOVER ARQ mode, only two stations can communicate at a time—just like AMTOR.) CLOVER shares another characteristic with AMTOR: the use of SELCALs. When attempting to contact another CLOVER station, you must send its SELCAL first.

What Do I Need to Run CLOVER?

The requirements for a CLOVER station differ substantially from those of a PacTOR station. They are:

❏ An SSB transceiver. The transceiver must be very stable (less than 30-Hz drift). The audio output from the CLOVER controller is fed to the audio input of the transceiver (CLOVER uses AFSK, not FSK). Receive audio is supplied to the controller from the external speaker jack or other source.

❏ An IBM-PC computer or compatible (80286 CPU or better).

❏ A CLOVER controller board. All CLOVER controllers are currently available from HAL Communications (see the RTTY/AMTOR Resource Guide for details). The HAL PCI-4000 CLOVER controller is installed *inside* the computer using any available expansion slot.

❏ HAL *PC-CLOVER* software. This is supplied by HAL Communications and is included with every PCI-4000 controller.

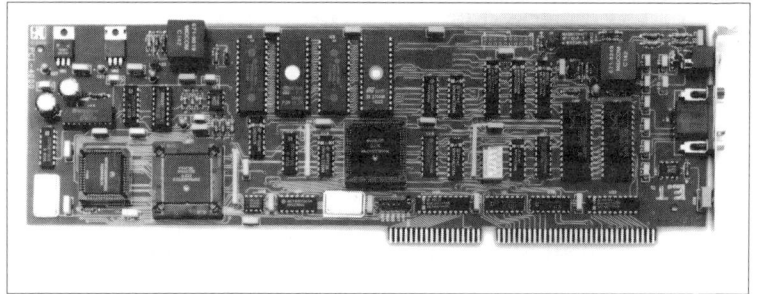

This is the HAL Communications PCI-4000 CLOVER controller. It fits conveniently inside your IBM PC or compatible computer. The only external wiring are audio cables to and from your SSB transceiver.

PacTOR or CLOVER? Which is Best?

The answer to this question depends on your interests. If you require the best possible HF digital communications, CLOVER is the clear winner. While PacTOR is fairly efficient, it doesn't come close to CLOVER.

On the other hand, CLOVER may require a greater equipment investment on your part. From the standpoint of economy, PacTOR looks very good.

RTTY and AMTOR have coexisted on the HF bands for over 10 years, each with its own devoted fans. I have a feeling that the same will be true of PacTOR and CLOVER. Both modes offer a dramatic leap forward in HF digital communications. Each is capable of sending and receiving error-free data (ASCII or binary) at excellent speeds under marginal conditions. It's a safe bet that PacTOR and/or CLOVER will be strong players in the future of HF digital communications!

CHAPTER 6
RTTY/AMTOR Contesting

When was the last time you enjoyed a good, clean fight? No, not the bare-knuckle variety! I mean a good-natured battle between you and your fellow digital enthusiasts on the HF bands. The goal of the conflict is simple: contact as many stations as possible during the time allotted.

Contesting is fun regardless of the mode, but it is especially exciting on RTTY and AMTOR. A good contest will sharpen your skills as an operator—and prove how long you can stare at a monitor before your body *demands* sleep! Contests are also excellent tests of your station equipment. It's fascinating to review your post-contest data and discover patterns. (Did most of your contacts seem to come from a particular area of the globe? Where was your most distant contact?)

If you're gunning for new states or countries to clinch an award, a contest is one of the best times to grab them. You can send a QSL card for a contest contact just as you would for a normal conversation. And if you're lucky enough—or skilled enough—to win in a contest category, you'll receive a handsome award.

David, WB0QIR, beat his local antenna restrictions by running coax out to the antenna on his car! He bagged a total of 153 contacts during the 1992 ARRL RTTY Roundup.

Don't fall into the trap of thinking that your station isn't up to the challenge of a contest. Brute force and racks of rigs don't necessarily create winners. The real key is *skill*. It's knowing, for example, when to hunt-and-pounce, or when to sit at one frequency and call "CQ Contest." It's watching the shifting propagation and checking various bands for new activity. In so many contests, it isn't the wealthy operator who wins, it's the *smart* operator!

Most HF digital contests involve the use of RTTY, AMTOR, ASCII or packet. At the time of this writing, PacTOR and CLOVER were not popular contest modes, but you can expect that to change very soon.

There are ten annual RTTY/AMTOR contests. Let's discuss each one in capsule form. For more detailed information, check *QST* magazine or the *RTTY Journal Contester's Guide* (see the RTTY/AMTOR Resource Guide).

Okay, so I only made 53 contacts during the 1992 ARRL RTTY Roundup. I still managed to earn this attractive certificate!

ARRL RTTY Roundup

The ARRL RTTY Roundup is one of the most popular digital contests of the year. It takes place on the first full weekend of January, beginning at 1800 UTC Saturday and ending at 2400 UTC Sunday. The eligible modes include RTTY, AMTOR, ASCII and Packet. (Packet contacts must be *direct*; no repeating devices allowed.) All bands from 80 through 10 meters can be used except 30, 17 and 12 meters.

US stations must exchange signal reports and the states where they are located; Canadian stations provide signal reports and provinces. DX stations must send signal reports and a serial number beginning with 001.

EA RTTY Contest

This contest is sponsored by the Seccion Territorial

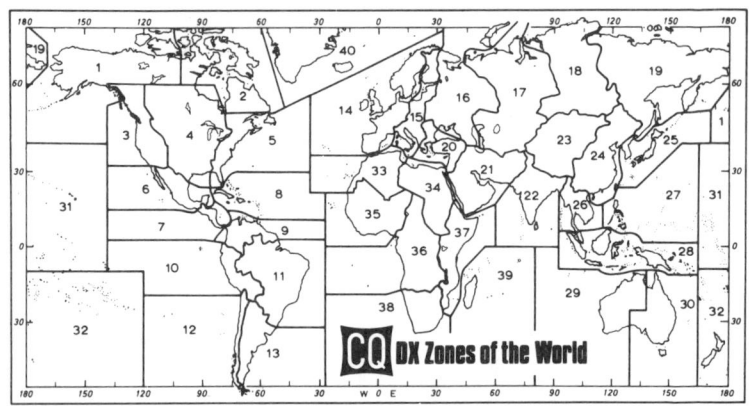

Fig 6-1—The CQ world zone map. Many contests require that you exchange your CQ zone along with other information.

Comarcal De Ure De Aranda De Duero in Spain. (Now you understand why everyone just calls it the EA RTTY Contest!) The contest gets underway on the first full weekend of February starting at 1600 UTC Saturday and ending at 1600 UTC Sunday. All bands except 30, 17 and 12 meters may be used.

Spanish stations must send signal reports, province prefix and CQ zone (see Fig 6-1). All others send signal reports and CQ zones.

BARTG RTTY Contest

BARTG stands for British Amateur Radio Teleprinter Group. Their contest begins at 0200 UTC Saturday on the third full weekend of March. It ends at 0200 UTC Monday. All bands except 30, 17 and 12 meters may be used.

Stations must exchange signal reports, the time in UTC (here's a good use for your automatic date/time function!) and a QSO number. Your first QSO is number 001. Your second is 002 and so on.

> **Catch a Glimpse of a RTTY Contest**
>
> What's it really like in the heat of a RTTY competition? See for yourself! The following is actual text copied during the 1992 CQ/RTTY Journal contest.
>
> **CQ CQ CQ CONTEST DE HB9CAL HB9CAL HB9CAL PSE K**
>
> **HB9CAL HB9CAL DE W9IT W9IT W9IT W9IT K**
>
> (W9IT tries to answer, but HB9CAL responds to another station instead)
>
> **GMØILB DE HB9CAL RR TNX . . . RST 599 ZONE 14 QSL? GMØILB DE HB9CAL PSE K**
>
> (W9IT tries again and succeeds)
>
> **HB9CAL HB9CAL DE W9IT W9IT W9IT K**
>
> **W9IT W9IT DE HB9CAL UR RST 599 599 ZONE 14 QSL? K**
>
> **THANK YOU . . . 599-ZONE 04-ILLINOIS . . . GOOD LUCK . . . HB9CAL DE W9IT SK**
>
> (Shifting to another frequency, we find N9ITX calling CQ)
>
> **CQ CQ CQ CONTEST DE N9ITX N9ITX K**
>
> **N9ITX DE W9JJX W9JJX UR 579 ZONE 04 INDIANA K**
>
> (N9ITX misses W9JJX's state and desperately asks for a repeat)
>
> **STATE STATE STATE AGAIN PSE DE N9ITX K**
>
> **N9ITX DE W9JJX STATE IS INDIANA INDIANA INDIANA OK? K**
>
> **W9JJX DE N9ITX . . . 599 . . . ZONE 04 IL QSL? K**
>
> **ROGER ROGER. TNX AND 73 DE W9JJX SK**

SARTG AMTOR Contest

The SARTG AMTOR Contest—sponsored by the Scandinavian Amateur Teleprinter Group—is probably the most popular AMTOR-only contest. You'll encounter it on the third full weekend of April from 0000 to 0800 UTC Saturday, 1600-2400 UTC Saturday and 0800 to 1600 UTC Sunday. (Nice of them to provide 8-hour rest breaks!) All bands are fair game except 30, 17 and 12 meters.

Stations contact each other initially using FEC (mode B), but *must* switch to ARQ (mode A) to exchange the contest information. The exchange consists of a signal report, name and QSO number starting with 001.

VOLTA RTTY DX Contest

The Como SSB/RTTY club and the Associazoione Radioamatori Italiani sponsor the VOLTA RTTY DX Contest to honor the Italian discoverer of electricity, Alessandro Volta. It takes place during the second full weekend of May

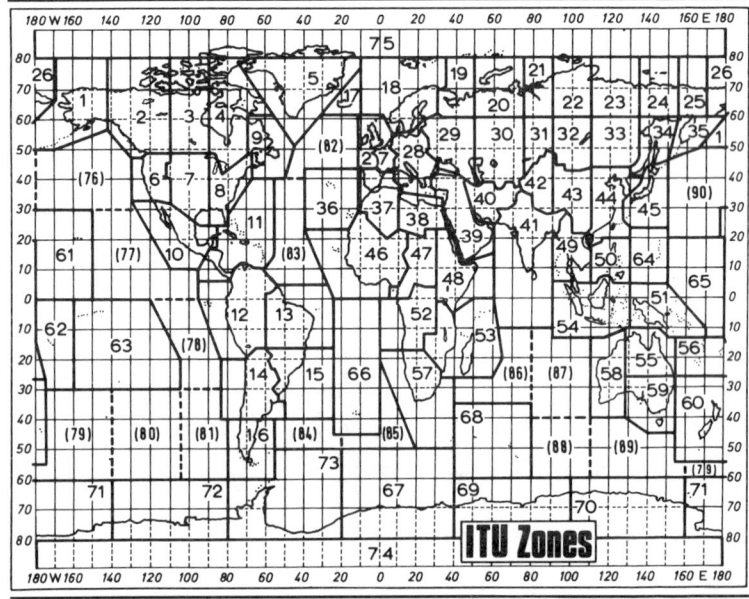

Fig 6-2—ITU zones of the world. In some competitions, your *ITU* zone is required rather than your CQ zone.

from 1200 UTC Saturday to 1200 UTC Sunday. All bands except 30, 17 and 12 meters may be used.

To score, the exchanges must include a signal report, QSO number (beginning with 001) and ITU zone number (see Fig 6-2).

ANARTS RTTY Contest

The Australian National Amateur Radio Teleprinter Society (ANARTS) sponsors a late spring contest that really brings out the DX activity. The contest starts at 0000 UTC Saturday on the second full weekend of June. It ends at 0000 UTC the following Monday. All amateur bands except 30, 17 and 12 meters are eligible and all digital modes including packet may be used (packet contacts must not use repeating devices).

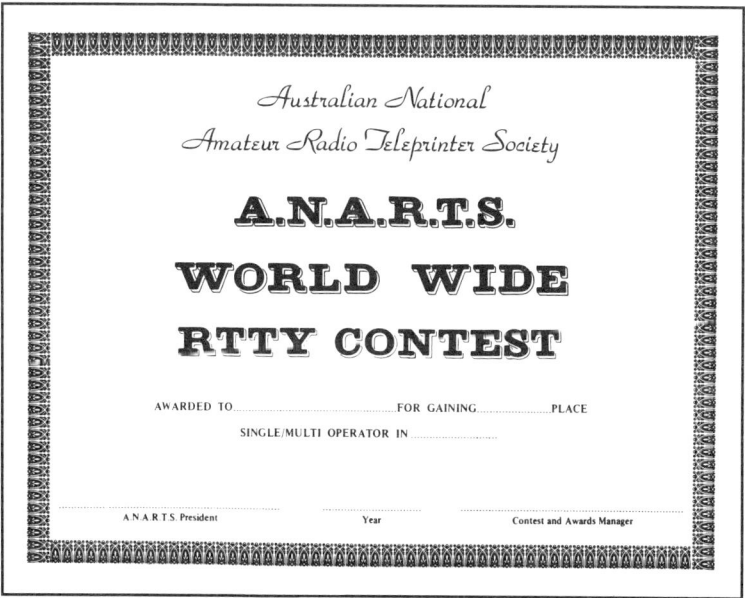

Depending on your final score, you might earn this award for participating in the ANARTS RTTY contest.

The official exchange is a signal report, time (in UTC) and ITU zone number.

SARTG RTTY Contest

The Scandinavian Amateur Teleprinter Group sponsors their second contest of the year—this one a RTTY contest—in late summer. The contest takes place on the third full weekend of August from 0000 to 0800 UTC Saturday, 1600 to 2400 UTC Saturday and 0800 to 1600 UTC Sunday. All bands except 30, 17 and 12 meters are eligible.

The exchange for this contest is very simple: a signal report and QSO number starting with 001.

CQ/RTTY Journal Contest

Cosponsored by *CQ Magazine* and the *RTTY Journal*, the CQ/RTTY Journal competition ranks with the ARRL RTTY Roundup as one of the most popular digital contests. Look for it in the last full weekend of September, beginning at 0000 UTC Saturday. The contest ends at 2400 UTC Sunday. All digital modes including packet may be used (no unattended operation or contacts through gateways or digipeaters). Every amateur band is eligible except 30, 17 and 12 meters.

Stations within the continental US and Canada exchange signal reports, state or province and CQ zone number (see Fig 6-1). All other stations send signal reports and CQ zones.

JARTS RTTY Contest

The Japanese Amateur Radio Teletype Society inaugurated the JARTS RTTY contest in 1992. The contest is held on the third weekend of October from 0000 UTC Saturday to 2400 UTC Sunday. This is a RTTY-only contest on all amateur bands except 30, 17 and 12 meters.

The contest exchange is a bit unusual: signal report and

operator age. JARTS rules specify that an age of "zero" is acceptable from female operators. (No kidding.)

WAE RTTY Contest

The final contest of the calendar year is the German WAE RTTY competition on the second week of November. It begins at 1200 UTC Saturday and ends at 2400 UTC Sunday. All bands except 30, 17 and 12 meters may be used.

The exchange is a straightforward signal report and QSO number beginning with 001.

RTTY Contest Software

You can streamline your contest operating—and increase your scores—by using specialized RTTY contest software. Many of these programs harness the power of your computer to provide handy features such as automatic contact logging. All you have to do is make the contacts and the computer will maintain your log!

The software packages listed below are primarily for IBM PCs and compatibles. Even so, contact the manufacturers to see if they market other versions for your computer.

ARIES
PO Box 830
Dandridge, TN 37725
tel 615-397-0742
Works with the KAM, PK-232 and HK-232 MCPs as well as standalone TUs

COMPRTTY II
David Rice, KC2HO
144 N Putt Corners Rd
New Paltz, NY 12561

tel 914-255-3273
Works with the KAM, PK-232, MFJ-1278 MCPs as well as the PCI-3000 and standalone TUs

SCOTCHLG
Hal Blegen, WA7EGA
2021 E Smythe Rd
Spangel, WA 99031
Works with the PK-232 MCP only

Wyvern Technology
Ray Ortgiesen, WF1B
35 Colvintown Rd
Coventry, RI 02816
tel 401-823-RTTY
Works with the KAM, PK-232 and MFJ-1278 MCPs

A RTTY/AMTOR Glossary

Terms in italics appear elsewhere in this Glossary.

ACK: An abbreviation for "acknowledgment." AMTOR stations exchange ACKs to verify that data has been received without errors.

AFSK: An abbreviation for "audio-frequency-shift keying." A method of digital transmission that is accomplished by varying the frequency of an audio tone applied to the microphone or auxiliary audio input of a transmitter.

AMTOR: An acronym for Amateur Teleprinting Over Radio. A popular method of digital communication on the HF bands.

ANSWERBACK: (AMTOR only) The programmable message that is sent when a receiving station receives a *WRU* (Who Are You?) command.

APLink: A bulletin board system (BBS) program created by Vic Poor, W5SMM. Messages and files may be accessed by either HF AMTOR or VHF packet-radio stations.

ARQ: Automatic repeat request. This is the error detection mode of AMTOR, PacTOR and CLOVER. ARQ signals are easily recognized by their chirp-chirp sound. An ARQ link can be established only between two stations.

ASCII: An acronym for American Standard Code for Information Interchange (usually pronounced *as-key*). A standard method of encoding data so it can be understood by many different computers.

autobaud: A routine used by some *MCPs* to automatically adapt to the serial data rate of a computer or *terminal*.

Baudot: A term commonly used among Amateur Radio RTTY operators to specify a 5-bit teletype code.

bit: The shortest data pulse used to make a RTTY character. The *Baudot* code uses 5 data bits. *ASCII* code uses 8 bits (7 data plus one parity bit).

bit/s: An abbreviation for "bits per second," a measurement of the rate at which data is transferred from one device to another.

capture: To save incoming data to a disk file for later use.

CD: Control delay (AMTOR/PacTOR). A time delay inserted by the *slave IRS* between the end of its reception of the *ISS* data "chirp" and the transmission of its *Control Signal*.

CLOVER: A modulation and data protocol created by Ray Petit, W7GHM. CLOVER uses PSK and ASK modulation on a pulsed tone sequence. CLOVER modulation levels are adaptive and data may be sent at rates 10 to 100 times that of AMTOR or HF packet radio. The CLOVER signal bandwidth (500 Hz at −50 dB) requires half the spectrum of AMTOR and a quarter that of HF packet radio.

Control Signal: The single character sent by the *IRS* to acknowledge (*ACK*) or not-acknowledge (*NAK*) the data sent by the *ISS*. The ISS repeats characters when a NAK (or no response) is received from the IRS.

data rate: The rate at which data is sent from one station to another. 45 baud is the standard RTTY data rate (speed = 60 WPM). 50 baud (66 WPM), 57 baud (75 WPM), and 75 baud (100 WPM) may also be used. 110 baud is the common data rate for HF ASCII RTTY.

data terminal: A device that allows a human operator to communicate with a *TU* or *MCP*.

digital signal processing: Using software rather than hardware to encode or decode digital signals for various modes.

download: The act of requesting and receiving specific data from another station.

DSP: An abbreviation for digital signal processing.

dumb terminal: A basic *data terminal* that provides only input and output functions. It cannot store or process data.

EIA-232-E: A data voltage and load protocol used by most computer devices for data pulses. An EIA-232-E *MARK* pulse has a negative voltage between –3 and –25 volts. A *SPACE* pulse is positive between +3 and +25 volts. Commonly referred to as *RS-232-C*.

End: The control command sent by an AMTOR station to end an *ARQ* link.

FEC: Forward error correction. A digital mode that may be used to send a message to more than one receiving station. Each character is sent twice to provide error correction. FEC is also called "Collective Broadcast" mode in commercial usage.

FIGS: A RTTY/AMTOR control character that signals the printer to shift to the FIGureS case.

firmware: Software stored in an integrated circuit memory chip.

floppy disk: Removable magnetic disks used to store digital data.

Forced Over: A command that can be initiated by the *IRS* to force a change in the channel direction (*IRS* to *ISS* and vice versa). The command used to cause a forced over varies with the controller or *MCP*.

FSK: Frequency-shift keying. Modulating a transmitter by using the data signal to shift the carrier frequency.

hard disk: Nonremovable magnetic disks used to store large amounts of data.

ISS: Information-sending station. The station that is sending information on the *ARQ* link. The ISS may be either the *master* or *slave* station.

IRS: Information-receiving station. The station that is receiving information on the *ARQ* link. The IRS may be either the *master* or *slave* station.

Listen: The mode of an MCP that allows it to monitor ongoing *ARQ* (and *FEC* and *SELCAL*) transmissions. Listen mode does not include error correction. Listen mode is generally not included in commercial *SITOR* controllers.

LTRS: A RTTY/AMTOR control character that signals a shift to the letters case.

MARK: The ON pulse state of a data signal. Also the "1" digital logic state.

Master: The station that establishes the *ARQ* link. The master station designation remains fixed for the duration of the ARQ QSO, regardless of which station is sending information. The master station sets the timing for both stations in an ARQ link.

MCP: An abbreviation for Multimode Communications Processor. A digital device designed to provide packet, CW, RTTY, fax, PacTOR and AMTOR communications as well as other modes. Also known as a multimode data controller.

Modem: A *MO*dulator-*DEM*odulator device that translates digital data pulses into audio tone frequencies or vice versa.

NAK: Not-acknowledge. The *IRS* response that tells the *ISS* station, "The last three characters were not received correctly. Please repeat last three characters."

NTS: National Traffic System. An ARRL-sponsored system for relaying messages throughout the nation and the world. NTS is supported by packet networks as well as CW, phone, RTTY and AMTOR.

Over: The control command that switches the roles of the two *ARQ* stations—*IRS* becomes *ISS* and ISS becomes IRS.

PacTOR: A modification of AMTOR and packet radio developed by DL6MAA and DF4KV. PacTOR uses the ASCII character code and adaptive data rate control to provide two to four times faster data throughput than AMTOR or HF packet radio.

PBBS: An abbreviation for Packet Bulletin Board System. A repository for packet mail, bulletins and other information within a local packet network.

port: A circuit that allows a one device (such as a computer) to communicate with another device (such as an *MCP*).

PTT: An abbreviation for Push To Talk. On a transceiver microphone, the button that's used to key the transmitter.

RAM: An acronym for Random Access Memory. A data storage device that can be written to and read from. Commonly used to refer to memory chips within a computer or other microprocessor-controlled device.

RS-232-C: The former standard for interfacing computers and/or *terminals* to other devices such as *TUs* and *MCPs*. It was replaced by *EIA-232-E*, although many still use the term RS-232 when referring to a "standard" interface.

RTTY: An acronym for Radioteletype (usually pronounced *ritty*). One of the oldest methods of digital communications. RTTY is still popular on the HF bands.

SELCAL: Selective call. The special letter (or number) sequence sent at the beginning of an AMTOR or CLOVER call in *ARQ* mode. An ARQ link will be established only when the transmitted SELCAL letters match those programmed at the desired station. Amateurs make up SELCAL codes as contractions of their amateur call signs; commercial ARQ stations are assigned numerical codes.

serial port: A circuit that permits one device to communicate with another by sending information bit-by-bit. Computers and *data terminals* often communicate with *TUs* and *MCPs* through their serial ports.

shift: The frequency difference between the *MARK* and *SPACE* pulses. The standard RTTY/AMTOR shift is 170 Hz.

SITOR: Simplex Teleprinting Over Radio. A commercial name for *ARQ* ship-to-shore communications.

Slave: The *ARQ* station that is called in an ARQ link. The slave designation remains fixed for the duration of the ARQ QSO. The slave always synchronizes its timing to that of the *Master* station.

SPACE: The OFF pulse state of a data signal. Also known as the "0" digital logic state.

split-frequency: Using a different frequency for transmitting and receiving.

Standby: The resting state of an AMTOR, PacTOR or CLOVER controller. When the programmed *SELCAL* or call sign is received, the controller automatically switches to *ARQ* mode and establishes the link. When called in

FEC mode, the controller automatically switches to FEC mode and prints the message.

start bit: The first bit sent in an asynchronous data transmission. The START bit is always a SPACE with a time duration equal to one data bit. For example, Baudot RTTY uses one start bit.

stop bit: The last data bit sent in an asynchronous data transmission. The stop bit is always a MARK pulse. With Baudot coding, the stop bit may be 1.41, 1.5, or 2 times the length of a data bit. In HF ASCII, the stop bit is two times the length of a data bit. At ASCII data rates greater than 110 baud, the stop bit is generally the same length as the data bit.

store and forward: The act of receiving data (messages or bulletins) and then storing them temporarily until they can be passed along to the next station.

switching delays: The delays associated with changing a transceiver or transmitter/receiver system from transmit to receive and back again.

TD: Transmit delay. A programmable time delay that blocks data transmission until the transmitter has reached full power.

terminal: A device that allows a human operator to communicate with an *MCP* or *TU*. Also known as a *data terminal*.

terminal software: Software that allows a personal computer to imitate (emulate) the functions of a *data terminal*.

terminal unit: A modem-type device dedicated to RTTY operating. It transforms digital data into *AFSK* tones or *FSK* pulses for transmission. A terminal unit also converts received audio to digital data. Also known as a *TU*.

throughput: A measure of the effectiveness of a data system. The throughput rate is the number of data elements per unit time (characters, bytes, or bits) that can be passed from one station to another without error. The throughput of AMTOR under ideal conditions is 6.67 characters per second (cps).

TTL: A type of digital interface based on transistor-transistor logic. For example, some personal computers offer a TTL interface rather than an EIA-232-E interface.

TU: See *terminal unit*.

upload: To send a data file or message to another Amateur Radio station.

WRU: Who Are You? A signaling system that lets one AMTOR controller automatically obtain identification from the other station. When enabled, reception of the FIGS-D ($) AMTOR code reverses the channel direction (over), sends text stored in the *ANSWERBACK* message, and restores channel direction (second over). WRU is often used by *APLink*.

RTTY/AMTOR Resource Guide

Books

The ARRL Operating Manual, 4th edition. American Radio Relay League, 225 Main St, Newington, CT 06111, tel 203-666-1541. Order No. 1086, $18

RTTY Contester's Guide, RTTY Journal, 1904 Carolton Lane, Fallbrook, CA 92028-4614, tel/fax 619-723-3838. $13

Newsletters

The RTTY Journal, 1904 Carolton Lane, Fallbrook, CA 92028-4614, tel/fax 619-723-3838. 10 issues/year. $15 US, Canada and Mexico; $30 elsewhere.

Equipment Manufacturers

Advanced Electronic Applications (AEA), PO Box C2160, Lynnwood, WA 98036-0918, tel 206-775-7373. (MCPs and TNCs)

HAL Communications Corp, PO Box 365, Urbana, IL 61801, tel 217-367-7373. (RTTY terminal units and CLOVER equipment)

Kantronics, 1202 E 23rd St, Lawrence, KS 66046, tel 913-842-7745. (MCPs and TNCs)

MFJ Enterprises, Box 494, Mississippi State, MS 39762, tel 800-647-1800. (MCPs and TNCs)

PacComm, 4413 N Hesperides St, Tampa, FL 33614-7618, tel 813-874-2980. (PacTOR equipment and TNCs)

RTTY/AMTOR Software Sources

Please note that the products and addresses shown below are subject to change. Unless otherwise noted, send a self-addressed envelope with 2 units of First Class postage when requesting information.

Apple (see also Macintosh)

Cotec, 13462 Hammons Ave, Saratoga, CA 95070.

Ham-Soft, PO Box 443, Galena Park, TX 77547-0443. Send $1 for catalog.

W1EO, 39 Longridge Rd, Carlisle, MA 01741.

Atari

ElectroSoft, 3413 N Duffield Ave, Loveland, CO 80538.

Langer, John, 115 Stedman St, #H, Chelmsford, MA 01824-1823, tel 508-256-6907.

Color Computer (Tandy)

Dynamic Electronics, PO Box 896, Hartselle, AL 35640, tel 205-773-2758.

Commodore

AMSOFT, PO Box 666, New Cumberland, PA 17070-0666, tel 717-938-8249. Catalog $1.

Colorburst, PO Box 3091, Nashua, NH 03061, tel 603-891-1588.

G & G Electronics, 8524 Dakota Dr, Gaithersburg, MD 20877, tel 301-258-7373.

Ham-Soft, PO Box 443, Galena Park, TX 77547-0443. Send $1 for catalog.

WB1FOL, 77 Wentworth St, Malden, MA 02148.

IBM PCs and compatibles

AMSOFT, PO Box 666, New Cumberland, PA 17070-0666, tel 717-938-8249. Catalog $1.

American Software, PO Box 509, Suite M16, Roseville, MI 48066-0509.

Colorburst, PO Box 3091, Nashua, NH 03061, tel 603-891-1588.

Comtech Research, 5220 Milton Rd, Custar, OH 43511.

Dynamic Electronics, PO Box 896, Hartselle, AL 35640, tel 205-773-2758.

Ham-Soft, PO Box 443, Galena Park, TX 77547-0443. Send $1 for catalog.

InterFlex Systems Design Corp, PO Box 6418, Laguna Niguel, CA 92607-6418.

Kasser, Joe, G3ZCZ, PO Box 3419, Silver Spring, MD 20918.

K-Quest, PO Box 92877, Southlake, TX 76092.

Renaissance Software & Development, Killen Plaza, Box 640, Killen, AL 35645, tel 205-757-5928.

Rice, David, 144 N Putt Corners Rd, New Paltz, NY 12561.

Schnedler Systems, 25 Eastwood Rd, PO Box 5964, Asheville, NC 28813, tel 704-274-4646.

Wyvern Technology, 35 Colvintown Rd, Coventry, RI 02816, tel 401-823-RTTY. (RTTY contest software)

Macintosh

Ham-Soft, PO Box 443, Galena Park, TX 77547-0443. (Send $1 for catalog.)

Krueger, Kevin, 1780 Ruth St, Maplewood, MN 55109, tel 612-770-0370. (RTTY contest software)

ZCo Corporation, PO Box 3720, Nashua, NH 03061, tel 603-888-7200.

Zihua Software, PO Box 51601, Pacific Grove, CA 93950, tel 408-372-0155.

Texas Instruments

Olson, Stuart, 6625 W Coolidge St, Phoenix, AZ 85033. (*Mass Transfer*, a general purpose terminal program)

About The American Radio Relay League

The seed for Amateur Radio was planted in the 1890s, when Guglielmo Marconi began his experiments in wireless telegraphy. Soon he was joined by dozens, then hundreds, of others who were enthusiastic about sending and receiving messages through the air—some with a commercial interest, but others solely out of a love for this new communications medium. The United States government began licensing Amateur Radio operators in 1912.

By 1914, there were thousands of Amateur Radio operators—hams—in the United States. Hiram Percy Maxim, a leading Hartford, Connecticut, inventor and industrialist saw the need for an organization to band together this fledgling group of radio experimenters. In May 1914 he founded the American Radio Relay League (ARRL) to meet that need.

Today ARRL, with about 160,000 members, is the largest organization of radio amateurs in the United States. The League is a not-for-profit organization that:

• promotes interest in Amateur Radio communications and experimentation

• represents US radio amateurs in legislative matters, and

• maintains fraternalism and a high standard of conduct among Amateur Radio operators.

At League Headquarters in the Hartford suburb of Newington, the staff helps serve the needs of members. ARRL is also International Secretariat for the International Amateur Radio Union, which is made up of similar societies in more than 100 countries around the world.

ARRL publishes the monthly journal *QST*, as well as newsletters and many publications covering all aspects of Amateur Radio. Its Headquarters station, W1AW, transmits bulletins of interest to radio amateurs and Morse Code practice sessions. The League also coordinates an extensive field organization, which includes volunteers who provide technical information for radio amateurs and public-service activities. ARRL also represents US amateurs with the Federal Communications Commission and other government agencies in the US and abroad.

Membership in ARRL means much more than receiving *QST* each month. In addition to the services already described, ARRL offers membership services on a personal level, such as the ARRL Volunteer Examiner Coordinator Program and a QSL bureau.

Full ARRL membership (available only to licensed radio amateurs) gives you a voice in how the affairs of the organization are governed. League policy is set by a Board of Directors (one from each of 15 Divisions). Each year, half of the ARRL Board of Directors stands for election by the full members they represent. The day-to-day operation of ARRL HQ is managed by an Executive Vice President and a Chief Financial Officer.

No matter what aspect of Amateur Radio attracts you, ARRL membership is relevant and important. There would be no Amateur Radio as we know it today were it not for the ARRL. We would be happy to welcome you as a member! (An Amateur Radio license is not required for Associate Membership.) For more information about ARRL and answers to any questions you may have about Amateur Radio, write or call:

ARRL Educational Activities Dept
225 Main Street
Newington, CT 06111
(203) 666-1541

Index

(Note: "RG" refers to RTTY/AMTOR Resource Guide.)

A

ACK (function of): 1-8
Amplifiers, RF power: 2-14
ANARTS RTTY Contest: 6-6
APLink: 4-14
   Common commands: 4-26
   NTS traffic: 4-27
   Using: 4-24
   Worldwide list: 4-17
ARQ: 4-2
   Operating: 4-2
   Propagation delay: 4-3
   Timing: 4-3
   Timing (adjustments): 4-6
ARRL Operating Manual: 1-5, RG-1
ARRL RTTY Roundup: 6-3
Art, RTTY: 3-21
Attenuator, audio: 2-20
Audio:
   Attenuator: 2-20
   Filters: 2-17
   Receive: 2-11
Autostart: 3-15

B

Bands, RTTY/AMTOR: 3-1
BARTG RTTY Contest: 6-4
Baudot, history of: 1-6
Brag tape: 3-9
Buffers, type-ahead: 3-7

C

Cables: 2-19
CLOVER: 5-9
   Bandwidth: 5-12
   Equipment requirements: ... 5-13
   FEC mode: 5-13
   Handshaking: 5-11
   vs. PacTOR: 5-14
Computers: 2-2
   Buying new: 2-4
   Buying used: 2-5
   Contest software: 6-9
   Terminal programs: ... 2-3, RG-2
Contests: 6-1
   ANARTS RTTY: 6-6
   ARRL RTTY Roundup: 6-3
   BARTG RTTY: 6-4
   CQ/RTTY Journal: 6-8
   EA RTTY: 6-3
   JARTS: 6-8
   Operating example: 6-5
   SARTG AMTOR: 6-5
   SARTG RTTY: 6-7
   Software: 6-9
   VOLTA RTTY: 6-6
   WAE RTTY: 6-9
CQ:
   AMTOR (answering): 4-14
   AMTOR (FEC): 4-8
   RTTY: 3-9
CQ/RTTY Journal Contest: ... 6-8

Index   135

D

Data terminals: 2-5
Date/time stamping: 3-10
Diddle: 3-4
Displays, tuning: 3-2
Dumb terminals: 2-5
DX:
   RTTY pileups: 3-11
   Split-frequency operating: . 3-13

E

EA RTTY Contest: 6-3
EIA-232E: 2-2
Equipment:
   CLOVER: 5-13
   Delays (AMTOR): 4-3
   Manufacturers (list): RG-1
   PacTOR: 5-8
   RTTY/AMTOR: 4-1

F

FEC: 4-1
FIGS: 1-6
Filters:
   Audio: 2-17
   IF: 2-16
Forced over (AMTOR): 4-11
Frequencies:
   APLink: 4-17
   Displays: 2-13
   MARK/SPACE tones: 1-10
   MCP shift: 1-12
   RTTY/AMTOR subbands: ... 3-1
   Stability: 2-12
   Suppressed carrier: 1-10
Frequency-Shift Keying (see FSK)
FSK: 1-5, 1-9
   Other shifts: 3-3
   vs. AFSK: 1-11, 2-11

G

Grounding: 2-19

I

IF filters: 2-16
Information receiving station
   (AMTOR): 4-2
Information sending station
   (AMTOR): 4-2
Inversion (MARK/SPACE): .. 3-3
IRS: 4-2
ISS: 4-2

J

JARTS RTTY Contest: 6-8

K

Keying (older rigs): 2-21

L

Listening to AMTOR: 4-7
Listening to RTTY: 3-3
LTRS: 1-6

M

Mailboxes, AMTOR (see APLink)
Mailboxes, PacTOR (see PacTOR)
Mailboxes, RTTY (see MSOs)
Manufacturers (list): RG-1
MARK: 1-5
Master (AMTOR): 4-10
MCPs:
   Frequency shift: 1-12
   Listening to RTTY: 3-3
   Tuning indicators: 3-2
   vs. TUs: 2-7
Memory ARQ (PacTOR): 5-8
Mode A: 4-2
Mode B: 4-1

Monitoring:
  AMTOR: 4-7
  RTTY: 3-3
MSOs:
  Autostart: 3-15
  Common commands: 3-17
  National Autostart
    Frequency: 3-16
  Unattended operation: 3-15

N

NAK (function of): 1-8
National Autostart Frequency
  (NAF): ...., ... 3-16
NTS (and APLink): 4-27

O

Operating:
  AMTOR ARQ: 4-2
  AMTOR FEC: 4-1
  Answering an AMTOR CQ: 4-14
  APLink: 4-14
  Bad habits: 3-10
  Brag tape: 3-9
  Calling CQ on AMTOR: 4-8
  Calling CQ on RTTY: 3-9
  Contests: 6-1
  Date/time stamping: 3-10
  Dropping the link
    (AMTOR): 4-11
  Forced over (AMTOR): 4-11
  Listening to AMTOR: 4-7
  Listening to RTTY: 3-3
  Master/slave (AMTOR): 4-10
  Over command (AMTOR): 4-11
  PacTOR: 5-4
  Prosigns: 3-5
  RST: 3-8
  RTTY: 3-6
  RTTY DX: 3-11
  RTTY pictures: 3-21
  RYRYRYRYRY: 3-10

SELCALs: 4-9
Sending stored files: 3-9
Split-frequency: 3-13
Tuning indicators: 3-2
Typing ahead: 3-7
Using split-screen software: . 3-7
WRU (AMTOR): 4-25
Over command (AMTOR): .. 4-11

P

PacTOR: 5-2
  Equipment requirements: 5-8
  Huffman code: 5-3
  Mailbox operation: 5-4
  Memory ARQ: 5-8
  Software: 5-10
  vs. CLOVER: 5-14
PC-CLOVER: 5-12
Pictures, RTTY: 3-21
Pileups, RTTY DX: 3-11
Port, serial: 2-2
Power:
  Amplifiers: 2-14
  Transceiver output: 2-12
Propagation delay (AMTOR): 4-3
Prosigns: 3-5
Publications:
  ARRL Operating Manual: ... 1-5,
    RG-1
  RTTY Journal: RG-1
  *RTTY Journal Contester's
    Guide*: 6-2, RG-1
  Your Packet Companion: 1-5

R

Radios (see Transceivers)
RF amplifiers (power): 2-14
RS-232: 2-2
RST: 3-8
RTTY Journal: RG-1
*RTTY Journal Contester's
  Guide*: 6-2, RG-1

RTTY/AMTOR, history of: ... 1-5
RYRYRYRYRYRY: 3-10

S

SARTG AMTOR Contest: 6-5
SARTG RTTY Contest: 6-7
SELCAL: 4-9
Serial port: 2-2
SITOR: 1-8
Slave (AMTOR): 4-10
Software:
  CLOVER: 5-12
  PacTOR: 5-10
  RTTY contest: 6-9
  Split-screen (description): 3-7
  Suppliers list: RG-2
  Terminal (description): 2-3
SPACE: 1-5
Split-frequency operating: ... 3-13
Split-screen software
  (description): 3-7
Station diagram: 2-1
Suppressed carrier frequency: 1-10

T

Tables:
  APLink Systems Worldwide:4-17
  Common APLink
    Commands: 4-26
  Common RTTY MSO
    Commands: 3-17
  RTTY/AMTOR subbands: ... 3-1
  Terminal software: 2-3
  Suppliers: RG-2
Terminal units (see TUs)
Terminals (data): 2-5
TOR: 1-8
Traffic, NTS: 4-27
Transceivers: 2-10
  Audio attenuator: 2-20
  Audio filters: 2-17
  Frequency displays: 2-13

Frequency stability: 2-12
FSK vs. AFSK: 2-11
IF filters: 2-16
Keying (older rigs): 2-21
Receive audio: 2-11
RF output: 2-12
Tuning indicator: 3-2
TUs:
  Description: 2-9
  Listening to RTTY: 3-3
  Tuning indicator: 3-2
  vs. MCPs: 2-7
Type-ahead buffers: 3-7

U

Upper/Lower case AMTOR: .. 1-8

V

VOLTA RTTY Contest: 6-6

W

WAE RTTY Contest: 6-9

YOUR
RTTY/AMTOR
COMPANION

PROOF OF
PURCHASE

ARRL MEMBERS

This proof of purchase may be used as a $0.80 credit on your next ARRL purchase. Limit one coupon per new membership, renewal or publication ordered from ARRL Headquarters. No other coupon may be used with this coupon. Validate by entering your membership number—the first 7 digits on your QST label—below:

FEEDBACK

Please use this form to give us your comments on this book and what you'd like to see in future editions.

License class:
☐ Novice ☐ Technician ☐ Technician with HF privileges
☐ General ☐ Advanced ☐ Extra

Name	ARRL member? ☐ Yes ☐ No
_____	Call sign _____
Daytime Phone ( ) _____	Age _____
Address _____
City, State/Province, ZIP/Postal Code_____
If licensed, how long? _____
Other hobbies _____

For ARRL use only
Edition 1 2 3 4 5 6 7 8 9 10 11 12
Printing 1 2 3 4 5 6 7 8 9 10 11 12

Occupation _____

From _____
_____
_____

> Please affix postage. Post Office will not deliver without postage.

EDITOR, YOUR RTTY / AMTOR COMPANION
AMERICAN RADIO RELAY LEAGUE
225 MAIN ST
NEWINGTON CT 06111

·· please fold and tape ··